Sampling and Monitoring of Environmental Contaminants

Richard C. Barth
Front Range Community College

Karl Topper
Mesa State College

McGraw-Hill, Inc.
College Custom Series

New York St. Louis San Francisco Auckland Bogotá
Caracas Lisbon London Madrid Mexico Milan Montreal
New Delhi Paris San Juan Singapore Sydney Tokyo Toronto

McGraw-Hill's **College Custom Series** consists of products that are produced from camera-ready copy. Peer review, class testing, and accuracy are primarily the responsibility of the author(s).

Sampling and Monitoring of Environmental Contaminants

Copyright © 1993 by McGraw-Hill, Inc. All rights reserved. Printed in the United States of America. Except as permitted under the United States Copyright Act of 1976, no part of this publication may be reproduced or distributed in any form or by any means, or stored in a data base retrieval system, without prior written permission of the publisher.

20 HAM HAM 0 9

ISBN 0-07-005153-4
ISBN-13: 978-0-07-005153-9
Editor: Mitchell W. Beaton

Cover Design: Maggie Lytle

Printer/Binder: Hamco/NetPub Corporation

PREFACE

This preliminary edition of "Sampling and Monitoring of Environmental Contaminants" is intended for use in a college level course. The format is intended to increase readability and allow students to take notes directly in the manual. Students we instruct seem to appreciate this format.

The manual focuses on a variety of environmental contaminants, particularly those commonly found at hazardous waste sites. Standard sampling protocols are used whenever possible, especially those promulgated by the Environmental Protection Agency. However, the manual is intended to be broad in scope and not tied to any agency's particular view on the details of sampling. Therefore, the student must apply this basic information to their particular sampling situation that, among other issues, must address any regulatory procedures that apply to their site.

This manual was first printed in preliminary form with the hope of receiving input from users, both instructors and students. We recognize the need for a unit dealing with sampling tanks and other containers, and the need for more detail and illustrations in most sections. We cordially solicit your suggestions for revisions.

The authors would like to thank Jim McLean of Golden, Colorado for illustrating this manual.

Richard C. Barth, Ph.D.
Front Range Community College
3645 West 112th Avenue
Westminster, Colorado 80030
Fax: (303) 466-1623

Karl Topper, Ph.D.
Mesa State College
P.O. Box 2647
Grand Junction, Colorado 81502
Fax: (303) 248-1324

August 1, 1993

Sampling and Monitoring of Environmental Contaminants

Table of Contents

UNIT 1 — SAMPLING BASICS

1.1	Sampling Objectives	2
1.2	Sampling Plan	3
1.3	Sampling Design	4
1.4	Representative Samples	7
1.5	Types of Samples	8
1.6	Sampling Equipment	9
1.7	Sampling Containers	9
1.8	Labels	12
1.9	Sample Preservation	12
1.10	Shipping	15
1.11	Chain-of-Custody	16
1.12	Decontamination	17
1.13	Field Notebook	21
1.14	Interfacing with the Laboratory	21
1.15	Quality Assurance/Quality Control	24
1.16	Safety Plan for Sampling	27
1.17	Sources of Information	28

(continued on next page)

(continued from previous page)

Table of Contents

Appendix A - Field Sampling Plan 31

Appendix B - Sampling Health & Safety Plan 42

UNIT 2 — SURFACE WATER

2.1	Regulatory Basis	58
2.2	Sampling Streams and Rivers	61
2.3	Sampling Lakes and Impoundments	67
2.4	Sampling Drinking Water	72
2.5	Sampling Storm Water Runoff	74
2.6	Sources of Information	77

Appendix 77

UNIT 3 — GROUND WATER SAMPLING

3.1	Federal Regulations Affecting Ground Water Monitoring	82
3.2	Ground Water Quality Standards	85
3.3	Characteristics of Ground Water	87
3.4	Ground Water Monitoring Wells	94
3.5	Well Development	120
3.6	Ground Water Sampling	126
3.7	Decontamination	148

(continued on next page)

Table of Contents

3.8	Quality Assurance and Quality Control	151
3.9	Sources of Information	152
Appendix A	Colorado Well Permit Form	154
Appendix B	Colorado Well Construction and Test Report	159
Appendix C	Forms Commonly Used in Well Development and Sampling	162
Appendix D	Colorado Well Closure Form	167

UNIT 4 — AIR SAMPLING

4.1	Regulatory Basis	170
4.2	Sampling Outdoor Air	171
4.3	Indoor Sampling	184
4.4	Sources of Information	193

UNIT 5 - SOLIDS SAMPLING

5.1	Regulatory Basis	194
5.2	Representative Sampling	195
5.3	Soil Characterization	196
5.4	Soil Sampling	203
5.5	Soil Sediments/Sludges	209

(continued from previous page)

Table of Contents

5.6	Sampling Sampling Bulk Materials	214
5.7	Sampling Artifacts	215
5.8	Sources of Information	215
Appendix		216

— Unit 1 —
Sampling Basics

Table of Contents

1.1	Sampling Objectives		2
1.2	Sampling Plan		3
1.3	Sampling Design		4
	1.3.1	Judgmental Sampling	4
	1.3.2	Random Sampling	5
	1.3.3	Systematic Sampling	6
	1.3.4	Combination Sampling Design (Hybrid Approach)	7
1.4	Representative Samples		7
1.5	Types of Samples		8
	1.5.1	Grab Sample	8
	1.5.2	Composite Sample	8
1.6	Sampling Equipment		9
1.7	Sampling Containers		9
	1.7.1	Container Size	10
	1.7.2	Material Compatibility	10
	1.7.3	Lids	11
	1.7.4	Container Contamination	11
1.8	Labels		12
1.9	Sample Preservation		12
1.10	Shipping		15
	1.10.1	Environmental Samples	15
	1.10.2	Environmentally Hazardous Substance Samples	16
1.11	Chain-of-Custody		16
1.12	Decontamination		17
1.13	Field Notebook		21
1.14	Interfacing with the Laboratory		21

(continued on next page)

(continued from previous page)

Table of Contents

      1.14.1   Selection of a Laboratory 22
      1.14.2   Containers 22
      1.14.3   Methods of Analysis 22
      1.14.4   Analytical Quality 23
      1.14.5   Dates 23
      1.14.6   Laboratory Safety 23

1.15   Quality Assurance/Quality Control 24

      1.15.1   Accuracy and Precision 24
      1.15.2   Quality Control Samples 24

1.16   Safety Plan for Sampling 27

      1.16.1   Purpose 27
      1.16.2   Hazard and Risk Identification 28
      1.16.3   Personnel Protection 28
      1.16.4   Decontamination 28
      1.16.5   Contingency Plan 28

1.17   Sources of Information 28

Appendix A - Field Sampling Plan 31

    A1.1   Background and Sampling Rationale 31

        A1.1.1   Background 31
        A1.1.2   Sampling Rationale 32

    A1.2   Phase I Investigation Program 32

        A1.2.1   IHSS 115 - Original Landfill 33

(continued on next page)

(continued from previous page)

Table of Contents

A1.3	Sample Analysis	37
	A1.3.1 Sample Designations	37
	A1.3.2 Analytical Requirements	38
	A1.3.3 Sample Containers and Preservation	39
	A1.3.4 Sample Handling and Documentation	39
	A1.3.5 Data Reporting Requirements	39
A1.4	Field QC Procedures	39

Appendix B - Sampling Health & Safety Plan 42

B1.1	Purpose and Policy	42
B1.2	Site Description and Scope-of-Work	42
	B.1.2.1 Background	42
	B.1.2.2 Potential Hazards	43
B1.3	Project Team Organization	43
B1.4	Health and Safety Risk Analysis and Medical Monitoring Requirements	44
B1.5	Emergency Response Plan and List of Emergency Contacts	44
	B1.5.1 Guidelines for Health and Safety Planning and Training	44
	B1.5.2 Emergency Recognition and Prevention	46
	B1.5.3 Personnel Roles, Lines of Authority and Communication Procedures During Emergency	47
	B1.5.4 Evacuation Routes and Procedures, Safe Distances and Places of Refuge	47
	B1.5.5 Decontamination of Personnel During an Emergency	47
	B1.5.6 Emergency Site Security and Control	48
	B1.5.7 Procedures for Emergency Medical Treatment and First Aid	48

(continued on next page)

(continued from previous page)

Table of Contents

B1.6		Required Levels of Protection and Air Monitoring	50
	B1.6.1	Personal Protective Equipment	50
	B1.6.2	Sampling Activity Equipment Needs	51
	B1.6.3	Heat Stress	52
	B1.6.4	Air Monitoring Procedures	53
B1.7		Site Control and Decontamination Procedures	53
	B1.7.1	Site Organization - Operation Zones	53
	B1.7.2	Exclusion Zone (Contamination Zone)	54
	B1.7.3	Contamination Reduction Zone	54
	B1.7.4	Support Zone	54
	B1.7.5	Decontamination Procedures	54

— Unit 1 —

SAMPLING BASICS

Prior to collecting samples, a substantial amount of work must be completed to organize all the details of the sampling program. Once in the field, the sampling process should read like a cookbook; all procedures are listed in a simple, sequential format. Obviously, the recipe, describing both the ingredients and the final product, is defined prior to sampling within a document generally referred to as a sampling plan. It is imperative the field samplers are thoroughly knowledgeable and trained in sampling procedures before collecting environmental or hazardous samples.

1.1 SAMPLING OBJECTIVES

Well-defined investigative objectives must be prepared prior to sampling. When preparing objectives, items often considered are:

- site characterization data;

- applicable regulations;

- nature and extent of suspected contamination;

- contaminant source characterization;

- potential receptors (soil, sediment, surface water, ground water, plants, animals, man, etc.);

- worker health and safety; and

- available resources.

The first stage in identifying the overall project objectives is essentially a preliminary planning stage in which the general investigative objectives are identified. This is followed by identifying the specific data quality objectives (DQOs). The DQOs identify necessary sampling objectives and the required level of data quality. Once this is clearly identified, then a sampling and analysis plan can be formulated.

1.2 SAMPLING PLAN

Prior to a sampling effort, a written sampling plan (Appendix A) should be prepared. The sampling plan should be integrated with analytical considerations so the document specifies the required analytical procedures and data quality levels. The importance of this relates to proper sample collection, preservation, and timely shipping; and to on-site analytical procedural requirements. The study objectives must be adequately considered with input from regulators, managers, samplers, chemists, etc., so the DQOs can be achieved and/or revised as necessary. Following are some of the items that should be included in a sampling plan:

- objectives (including DQOs);
- applicable regulations;
- site background information;
- suspected level of contamination;
- training of sampling team;
- sampling/field analysis methods (SOPs);
- justification of selected methods and procedures;
- organization of the investigative team;
- transportation & shipping information (e.g. chain of custody, shipping method);
- field notes;
- safety plan (e.g. decontamination; work zones);
- quality assurance/quality control;
- chain of custody;
- shipping;
- lab analytical protocol (parameters, methods, detection limits);
- interfacing with laboratory.

The standard operating procedures (SOPs) provide the recipe for performing a specific sampling or measurement operation. It includes items such as the instrumental calibration procedure, the required sampling equipment and proper use, sampling containers, sample preservation, etc.

Deviations from the sampling plan and SOPs should be rare and approved by the sampling director. Field sampling is not a place where extemporaneous changes, improvisation, or substitution are appropriate. However, unanticipated field conditions can lead to inappropriate specifications in the sampling plan. It is imperative that the field investigative team possesses sufficient knowledge so problems in the field can be identified and appropriately handled. Any deviations from the sampling protocol must be fully documented and justified.

1.3 SAMPLING DESIGN

Sampling design, or the selection of sampling locations, is critical to the success of the investigative program. Sampling design will determine the extent of sampling, how the data can be statistically or qualitatively analyzed, what conclusions can be supported from the data, and how well the data can be defended in court.

Most sampling designs can be categorized within the following sampling schemes: judgmental, random, systematic, and combination (hybrid approach).

1.3.1 Judgmental Sampling

This sampling design is often used when specific information is known about the configuration of the release. Specific information may include point of chemical release, direction of ground water flow, stained soil, air dispersal patterns, etc. This information is used to select the specific sampling location.

Obviously, this sampling design produces a bias toward the worst contamination, or a "worst case scenario." Additionally, areas of significant contamination can be missed. Sampling locations are selected on the basis of where the project manager or scientist thinks the contamination should be considering site characteristics and visual observations. Conversely, selection of "background" samples generally involves some type of judgment. Background samples (samples collected from an area presumably non-contaminated yet typical of the study area) are often collected for comparison to the presumed contaminated area. Such judgments are often wrong. Judgmental sampling cannot be analyzed statistically, thus severely limiting data interpretation and its defensibility.

However, this type of sampling is efficient and economical both in terms of sampling time and laboratory costs. It is probably the most common type of hazardous substance sampling and is often used during initial investigations to determine if contamination has occurred.

1.3.2 Random Sampling

In random sampling, every portion of the area under investigation has an equal chance of being selected for sample collection. For example, a researcher wants to know the average weight of apples from a particular tree. All the apples are picked, mixed thoroughly in a basket, and a blindfolded sampler picks out a certain number of apples. These apples are then weighted and the average weight calculated. The sample collected and the resulting data are free of bias.

Commonly applied classical statistics are based on the assumption that numerical data was collected at random, that the individual samples are independent of each other, and the sample population represents a normal distribution and is representative of the parent population. Strictly speaking, violation of any one of these assumptions invalidates the statistical analysis. However, this view is theoretical in nature and in practice, minor departures from strict randomness are allowed as long as the deviation does not introduce or allow any significant bias.

A random sampling design does not guarantee adequate spatial representativeness of the sampling area; all samples could be clustered in one corner. To address this problem, the number of samples collected could be increased and/or a larger sample size collected. A random design precludes investigation of areas where professional judgment suggests some significant spatial event worthy of sampling. In addition, sample locations may fall in areas that cannot be sampled. For example, when collecting soil samples, a random design could locate samples on top of drums, in paved areas, and in other locations where no soil exists. Finally, locating the sampling sites in the field may be difficult and time consuming because the locations must be absolutely bias free.

Randomization is typically applied to only one level of a study. For example, the surface locations of soil samples are found using a random design, but sample width and depth are kept constant. Randomizing at all levels may introduce unwanted variation and may invalidate the data collected.

Many experts feel that sampling for legal purposes often requires the collection of random samples so that all traces of bias are removed. Others will argue that legal defensibility relates primarily to the analytical program so field sampling should focus on adequate coverage rather than randomizing sample locations.

It is likely this debate will continue for some time.

1.3.3 Systematic Sampling

In a systematic sampling design, a grid is placed over a map of the study area or a grid is surveyed on the ground. The intersection of grid lines defines where a particular sample is to be collected. This sampling design is used in a variety of situations, such as defining the extent of soil contamination from a leaking underground storage tank. In such cases, traditional statistics may not be appropriate for data evaluation. However, geostatistical analysis is well suited for systematic sampling and for evaluation of spatially related sampling points.

The grid area is selected based on sampling needs. Thus, when soil sampling points are located in a paved area (or other impossible locations) they are eliminated. The shape of the grid may be square, rectangular, triangular, or radial, depending on the needs of the investigation. Spacing of the grid is based on the spatial distribution of the contamination. If the contamination is expected to be distributed in a homogeneous manner, wider spacing may be appropriate because little variation is expected among the samples. In contrast, a heterogenous distribution will require small spacing for adequate characterization. Distinction of the degree of spatial heterogeneity is difficult. The best approach is to sample a small area at an intensive sampling interval (e.g. 50 feet) and complete a semivariogram of the data. This type of geostatistical plot will determine the degree of spatial distribution a particular sample represents.

The advantage to systemic sampling is that all portions of the study site are sampled. In addition, impossible locations are avoided, and sampling and laboratory costs are not as high as in random sampling. Bias is limited to selection of the study area to be included in the sampling grid (which is actually a source of bias for all sampling procedures), shape of the grid, and spacing of the grid.

The prime disadvantage to systematic sampling is that it may involve a variable degree of judgment that could be challenged by regulators, other investigators, and attorneys. However, when properly designed and statistically evaluated using geostatistics, the subjectiveness of such judgement is removed and is fully defensible.

NOTES

1.3.4 Combination Sampling Design (Hybrid Approach)

Most sampling schemes consist of a combination of the types previously discussed. As an example, when selecting an appropriate sampling scheme for drums suspected of containing hazardous liquids, the drums might initially be staged and categorized according to preliminary information concerning their labeled contents, material (steel or plastic), and size, and then sampled randomly within the specified population groups. Stratified sampling involves division of the sample population into groups based upon knowledge of sample characteristics. This reduces the sampling area into a number of individual sampling units which are then randomly sampled within each subunit. The primary purpose of stratified sampling is to increase the precision of these divisions.

Another similar design entails combining a grid sampling approach with random sampling. This approach, when properly designed, tends to combine the advantages of both random and systematic sampling while avoiding most of the disadvantages. It allows some professional judgement to enter the decision of where to sample, but judgement is limited and significant bias is not introduced. An example of this approach would be to place a grid over a sampling area and randomly select sampling points within the boundaries of each set of grid lines.

A properly designed randomized grid eliminates most of the problems associated with judgmental and random sampling. Although the "complete randomness" basis of statistics is not strictly followed, the magnitude of deviation is usually not significant.

No matter which sampling scheme is utilized during a site assessment, one must maintain complete documentation of all decisions made in designing and implementing the sampling plan, and all actions must be justified.

1.4 REPRESENTATIVE SAMPLES

Regardless of the sample type, all samples must accurately depict or represent the substance being collected. For example, stagnant water in a ground water monitoring well does not accurately reflect the water flowing through the aquifer. Therefore, this stagnant ground water must be replaced with fresh ground water. Standard ground water sampling practice states that a certain amount of water (usually three to five casing volumes) be removed prior to sampling, or that water be removed until certain parameters (such as total dissolved solids, conductivity, and/or pH) are constant between samples. Once the stagnant water has been removed, a representative sample can be collected.

A more formalized statistical definition of representativeness is as follows: the degree to which the data accurately and precisely represent a characteristic of a population parameter, variation of a property, a process characteristic, or an operational condition.

1.5 TYPES OF SAMPLES

There are two basic types of samples: grab and composite. The type of sample collected is determined by sampling objectives and the nature of the material being sampled.

1.5.1 Grab Sample

A grab sample is an individual sample taken at a specific location and at a specific time. For example, the soil sampled at a specific location downgradient of a leaking storage tank is grab sampled. The use of the word "grab" is unfortunate, for it can imply that a handful of material is pulled from any convenient point. Grab samples must be collected according to the sampling plan that considers size, shape, location, representativeness, etc. of the sample. As the media characteristics vary over time and distance, the representativeness of grab samples will decrease.

1.5.2 Composite Sample

A composite sample is a combination of more than one sample collected at various sampling locations and/or different points in time. This mixing of samples is conducted to increase the representativeness of the sample and/or to obtain sufficient material for laboratory analysis. The data which results from a composite sample represents an average value which can be more cost effective than analyzing a number of individual grab samples and calculating an average value. However, the composite sample does not allow for evaluation of the variability inherent within the composite and will not provide the discrete location of contamination.

Compositing may be useful in determining the overall extent of a contaminated area, but should not be used as a substitute for characterization of individual constituent concentrations. Composite samples are of limited value because they do not reflect actual concentrations and can reduce some concentrations to below the detection limit. Therefore, compositing should be limited and should always be done in conjunction with an adequate number of grab samples.

Grab sampling techniques are generally preferred when sampling hazardous wastes due to the above-mentioned limitations of composite sampling. Additionally, grab sampling minimizes the samplers contact time with a concentrated waste, reduces risks associated with compositing unknowns (e.g. chemical incompatibility concerns), and eliminates chemical changes that might occur due to compositing.

1.6 SAMPLING EQUIPMENT

Regardless of the media to be sampled, equipment selection must consider the following items.

- **Safety.** The equipment selected must minimize both chemical and physical risks to the sampler.

- **Compatibility.** The equipment must be chemically compatible with the material being sampled. For example, metal sampling tools may not be appropriate for sampling corrosive wastes or when testing for metals.

- **Representativeness.** The sampling equipment must allow for the collection of a representative sample.

- **Ease of Use.** Complicated sampling equipment is prone to operator error and breakdown.

- **Decontamination.** The ability to completely decontaminate sampling equipment in the field is another criterion important in equipment selection.

1.7 SAMPLING CONTAINERS

A variety of factors effect the selection of sampling containers. Size, material compatibility, and lid type are among the more important factors. Selection of the appropriate container is generally based on the constituents being analyzed and the sample matrix (soil, water, air, sludge). Specifications regarding the types of sampling containers required must be completed prior to field sampling activities. The container type is often specified by the regulatory agency.

1.7.1 Container Size

Laboratory analysis requires a certain sample volume or weight (mass) for analysis. Therefore, a container of the appropriate size must be selected. Container sizes generally range from 20 to 1000 ml.

1.7.2 Material Compatibility

The sample must be compatible with the container; any type of sample-container reaction will destroy the integrity of the sample. General compatibility rules are as follows:

- glass is usually required when samples are to be analyzed for organics;

- plastic or glass can be used when samples are to be analyzed for metals (plastic is preferred);

- plastic should be used when samples are to be analyzed for boron or silica;

- amber glass should be use when samples are photosensitive;

- special volatile organic analysis (VOA) vials must be used when volatile organics in water are to be analyzed;

- glass should be used for solids and sludges requiring pH analysis; and

- glass is required for analysis of oil, grease, and phenolics.

Laboratories often supply the proper containers for a sampling program.

1.7.3 Lids

Lids are part of the sample container and must be considered, especially in terms of compatibility. The interior of the lid must be compatible with the sample and the analytical parameters. Most lids are lined with either Teflon or aluminum foil. In addition, some lids are specially designed with a Teflon insert (or septum) that allows analytical access to the sample without removing the cap.

Lids are available in either narrow or wide mouth forms. Narrow mouth lids are generally used for liquids because there is less likelihood of leakage. Wide mouth lids are generally used for sludges and solids.

1.7.4 Container Contamination

Containers must be free of any contaminants before they are used in the field. Sampling plans usually specify that certain cleaning procedures be used by the laboratory or other suppliers of sampling containers. The lot number and/or certificate of cleanliness provided by the container supplier should be requested.

Once the containers have been cleaned, additional "cleaning" in the field is usually unnecessary. For example, glass bottles are often cleaned by the supplier or laboratory using the following procedure:

- Bottles, and caps are washed in laboratory-grade, nonphosphate detergent.
- Rinsed 3 times with distilled water.
- Rinsed with 1:1 nitric acid.
- Rinsed 3 times with ASTM Type 1 organic-free water.
- Oven-dried for 1 hour.
- Rinsed with hexane.
- Oven-dried for 1 hour.

Tape should not be used to secure lids. Most tapes have organic glues that can penetrate the container cap and contaminate the sample. A properly secured lid does not need any additional security to prevent leakage.

Contamination is also minimized by not opening the sample

container until immediately prior to sample collection and by closing the container immediately after sample collection. Be sure to place the sample in a refrigerated cooler as soon as possible.

1.8 LABELS

All sampling containers must have labels. Labels are generally supplied by the laboratory and require a variety of information including company, project, sample identification such as location or number, sample date and time sampled, sample type (raw or filtered), and preservatives added (Figure 1).

Information required on the label should be written prior to sample collection because sample collection may result in a wet or dirty label that repels ink. The information on the label should agree with all entries on other forms, such as the chain-of-custody form and sample log sheet. Use black, waterproof pens to write on labels.

Most labels are of the self-adhesive type, and some are color-coded (such as red for metals, green for oil and grease, orange for organics, etc.). Labels are usually applied to the sample container prior to sample collection. Labels do not stick well to wet or dirty containers. However, sample material can react with and/or stain labels and render them illegible. If this potential exists, apply the label after sample collection.

Some labs will apply a color code or a computer bar code to the sample container. Such codes are used to speed processing time within the lab. Do not remove these codes.

1.9 SAMPLE PRESERVATION

The objective of sample preservation is to ensure that the chemical characteristics of the sample do not change from the time of collection to the time of analysis. Preservation techniques depend on the analytical parameters, and in most cases, the analytical laboratory will specify the required sample preservation procedures. Common preservation techniques are as follows.

- ♦ **Low Temperature**—immediately cooling samples to 4 degrees Centigrade is the basic sample preservation technique. This is implemented by using coolers and packaged coolants. However,

DATA CHEM
LABORATORIES

960 West LeVoy Drive
Salt Lake City, Utah 84123-2547
(801) 266-7700

4388 Glendale-Milford Road
Cincinnati, Ohio 45242-3706
(513) 733-5336

SITE IDENTIFICATION	DATE
ANALYSIS REQUESTED	TIME
	PRESERVATIVE
COMMENTS	

SPECIALLY CLEANED CONTAINERS
A SORENSON COMPANY

VISTA Laboratories Inc.
3830 High Court
Wheat Ridge, CO 80033
(303) 467-0630

COMPANY:		VISTA ID:
SAMPLE ID:		
DATE: / /	TIME: AM PM	SAMPLER'S INIT:
PARAMETERS:		
SAMPLE TYPE: ☐ RAW ☐ FILTERED	PRESERVATIVE: ☐ HCL ☐ H_2SO_4 ☐ ZnAc ☐ HNO_3 ☐ NaOH ☐ $Na_2S_2O_3$	

Evergreen Analytical Inc.
4036 Youngfield
Wheat Ridge, Colorado 80033
(303) 425-6021
FAX (303) 425-6854

SAMPLE IDENTIFICATION ______
LOCATION ______
DATE ______ TIME ______
SAMPLED BY ______
TEST (S) ______
PRESERVATIVE ______

EAGLE EP PICHER
ENVIRONMENTAL SERVICES
36 B.J. TUNNELL BLVD, EAST-MIAMI, OK 74354-3300
1-800-331-7425

Specially Cleaned Sample Container
LOT NO.:

DATE:	TIME:	COLLECTED BY:
SAMPLING SITE:		
SAMPLE TYPE ☐ Grab ☐ Composite ☐ Other ______		
TESTS REQUIRED:		PRESERVATIVE

ANALYTICA
—I-N-C-O-R-P-O-R-A-T-E-D—
18000 W. Highway 72
Golden, Colorado 80403 • (303) 420-4449

Client:
Project:
Location:
Date / Time:
Sampler:
Field Filtered: YES NO
Preservative:

Figure 1 Examples of labels

samples should not be allowed to freeze.

- **Acid Preservation**—liquid samples can be preserved by adding acid, usually nitric, sulfuric, or hydrochloric, to lower the pH to less than 2.

- **Basic Preservation**—liquid samples can be preserved by adding bases, such as sodium hydroxide, to increase the pH to greater than 9.

- **Filtering**—water samples that require analysis for dissolved metals must be field filtered. Consult with the client or regulatory agency before filtering. Generally, if a filtered sample is required, an unfiltered sample is also collected to determine total metal concentrations.

- **Other chemicals**—zinc acetate, sodium thiosulfate, and mercuric chloride are also used as preservatives for specific analyses.

Other preservation techniques include storage in a dark area (usually in a cooler), use of amber glass, and the elimination of head space in the filled container.

In some cases, the analytical laboratory will place the appropriate preservative in the sample containers prior to shipping the containers to the field. In other cases, preservatives are added in the field. This increases sampling flexibility in the field, prevents the concentrated preservative from reacting with the cap, and allows for preservatives to be reduced or eliminated if they react with the sample. However, handling preservatives in the field can be hazardous and increase the likelihood of field contamination.

Once a sample has been properly preserved, it has a shelf life or holding time. Holding times range from zero (such as pH for a liquid or dissolved oxygen) to months. Holding time is defined as the duration of time in which there will be no significant deterioration in sample quality. Regardless of the holding time, samples should be transported to the laboratory immediately after collection. The laboratory then analyzes the sample within the appropriate period of time. Regardless of the holding time, laboratory analysis should be completed as soon as possible. For example, volatile organic compounds (VOCs) can experience significant sample loss within hours after sample collection even though the accepted holding time ranges from 7 to 14 days.

Careful management of sample retention and holding times is critical when sample analysis is phased or triggered (second tier analysis depends on the results from the first tier).

1.10 SHIPPING

Prior to transporting samples to the laboratory, the containers must be properly packed to prevent breakage. Plastic bubble pack, styrofoam, or vermiculite are commonly used for cushioning. Vermiculite is probably the best packing material because it not only cushions but it absorbs any spilled liquid. However, this material is messy during unpacking, and some laboratories will not accept vermiculite.

Samples are normally packed in coolers. Custody seals, packing slips, chain-of custody forms, and other documents must accompany the shipment. In addition, the Department of Transportation (DOT) has specified packaging and labeling requirements for hazardous materials. In response to these and other shipping regulations, the EPA recommends the following sample packaging procedures for non-regulated "environmental samples" and DOT regulated "environmentally hazardous substance" samples.

1.10.1 Environmental Samples

The EPA recommends that the following procedures be followed for shipping environmental samples that are not under DOT regulations.

> Sample containers are labeled, sealed, and then sealed in a plastic bag.
>
> Containers are placed in a cooler with vermiculite pillows and cooling material.
>
> Chain-of-custody and other forms are secured inside the cooler lid.
>
> The cooler is taped shut with custody seals.
>
> Samples are shipped by the fastest available method.

1.10.2 Environmentally Hazardous Substance Samples

Depending on the volume and toxicity, some environmental samples, such as those containing hexane, phosphorous, radionuclides, and PCB's, are under DOT regulations when shipped. Generalized shipping procedures are as follows.

> Sample containers are labeled, sealed, and sealed in a plastic bag.
>
> Containers are placed in a sealed metal can with vermiculite pillows between the sample and the wall of the metal can.
>
> The metal can is sealed with tape.
>
> The metal can is placed in a cooler with cooling material, or in another appropriate container if cooling is not needed.
>
> Chain-of-custody is secured inside the cooler or packing container.
>
> Custody seals are placed on the cooler or other packing containers.
>
> Include the shippers certification with the container.
>
> Apply appropriate DOT labels to the outside of the shipping container.
>
> Ship the container by the fastest method available.

1.11 CHAIN-OF-CUSTODY

Samples collected during a site assessment may be used as evidence in court. Therefore, possession must be traceable from the time the samples are collected until the data are introduced as evidence during the legal proceedings. To document sample possession, chain-of-custody procedures must be followed.

The chain-of-custody form documents possession and sample control. The EPA defines possession (or custody) as follows:

- ♦ physical possession of samples;

- previous physical possession and samples are now in sight;

- previous physical possession and samples have been placed in a secure area (including a shipper); or

- previous physical possession and samples are now in a locked container.

Therefore, the field sampler is personally responsible for the care and custody of the samples collected until they are properly transferred to another person who accepts custody. Whenever possession changes, the chain-of-custody form (Figure 2) must be signed and dated by the person relinquishing custody and the new person accepting custody. The chain-of-custody form is wrapped in a plastic bag and placed inside the sample cooler.

Pressure-sensitive custody seals are usually placed over the lid of the cooler in such a manner that they are torn when the cooler lid is opened. In some sampling situations, custody seals are also placed over the lids of individual sampling containers. It is standard EPA protocol to use a sample custody seal in which the seal is placed over the container lid and down each side of the container in such a manner that opening the container will break the seal. However, the seals are not placed over the top of a septum cap because of potential contamination from the organic glue. If a custody seal is required for septa containers, place the seal around the bottom edge of the hard plastic cap.

The chain-of-custody also serves as a sample log sheet where sample description and analytical parameters are specified. As shown in Figure 2, a variety of other information may be required on this form. In addition, pertinent field observations should be written on the form. For example, a sample that has detectable vapors or unusual color should be noted.

1.12 DECONTAMINATION

Sampling equipment must be decontaminated (decon) between each sample to protect against cross contamination. Decontamination usually consists of washing equipment with a water-based solution (such as soap and water) followed by a rinse

C 30 (6)

CenrefLabs 695 North Seventh Street
Brighton, Colorado 80601-1559
Telephone 303 659-0497

CHAIN OF CUSTODY RECORD

COMPANY NAME ____________ CONTACT NAME: ____________ TELEPHONE NO. ____________

Samplers: Print Name		Signature									ANALYSIS								
Project No:						Sample Type													
SAMPLE ID	SAMPLE LOCATION	Composite	Grab	Date	Time	Water	Liquid	Oil	Solid	No. of Containers									REMARKS

COMMENTS: ____________

Relinquished by: (Signature)			Received by: (Signature)	Date	Time
Relinquished by: (Signature)			Received by: (Signature)	Date	Time
Dispatched by: (Signature)	Date	Time	Received for Laboratory by:	Date	Time
Method of Shipment:					

WHITE COPY—CENREF YELLOW COPY—CUSTOMER

Figure 2 An example of a chain-of-custody form.

CENREF LABS CHAIN-OF-CUSTODY INSTRUCTIONS

A sample is physical evidence collected from a specific location or from the environment. The following is a list of suggestions for completion of Cenref Chain-of-Custody records. Please reference Figure 4. A copy of the chain-of-custody record will be included in the final report.

(A) "Company Name" — You should always make sure your company name appears on each chain-of-custody.

(B) "Contact Name" — Include contact name, it is helpful to the laboratory in case we need to contact someone regarding the project.

(C) "Telephone Number" — Also helpful to the laboratory in contacting the client if needed.

(D) "Samplers Name & Signature" — The sampler should print and sign his name on all chain-of-custody sheets.

(E) "Project Number" — Include any information here that will help you to identify this project from your other projects.

(F) The sample information should be filled out as completely as possible, this information includes:

1. "Sample ID" — The identification you are giving each sample.
2. "Sample Location" — The location at which each sample was taken.
3. "Comp" or Grab" — Indicates whether this is a composite sample or a grab sample.
4. "Date" — The date sampled.
5. "Time" — The time sampled.
6. "Sample Type" — Indicates if the sample is a water, liquid, oil or a solid sample.

(G) "No. of Containers" — Number of containers submitted for each sample.

(H) "Analysis" — Include here all analyses requested for each sample. Use the boxes under the analyses requested to indicate which samples need which analyses.

(I) "Remarks" — Is there any additional information that would be helpful to either Cenref Labs, or useful to write a project report later. An example remark would be, "Sample has two phases, analyze water phase only."

(J) "Comments" — Any information you feel will be beneficial to Cenref Labs regarding this project such as, "Samples may have high concentration levels."

(K) "Relinquished By" — Should be signed by the person who either hand delivers the samples to the lab, or by the person who packs the samples for shipment to the lab.

(L) "Dispatch By" — If the sampler is not the person who delivers or ships the samples, the sampler should relinquish the samples to the deliverer by signing this section.

(M) "Method of Shipment" — This indicates how the sample was shipped to the laboratory.

(N) "Received By" — When samples have changed hands, the person receiving the samples needs to sign, date, and show the time the samples were received.

A custody seal is also sent with the samples. This seal should be attached to the outside of the cooler between the lid and body of the cooler. This helps ensure the integrity of the cooler.

Figure 3 Chain-of-custody instructions for the form shown in Figure 2.

solution (generally deionized/distilled water). However, specific decontamination procedures must be developed for each site. Decontamination solutions are based on the characteristics of the chemicals to be removed from the equipment and requirements of applicable regulations. Table 1 illustrates some of the commonly followed sequences for decon. Solutions are often dispensed in a pressurized portable tank, such as those commonly used to apply yard pesticides. Decon can involve flushing equipment with solvents, use of pressure or steam jets, heating, flaming or baking items, scraping, rubbing, grinding, etc.

All decon solutions must be collected and treated as hazardous waste if it is in contact with a hazardous material. In most cases, the solutions are collected in a drum and left on-site. However, the disposition requirements will depend greatly upon the applicable regulations of the specific site.

In addition to equipment decontamination, worker decontamination is necessary. Glove decontamination (or disposal) is especially important in that gloved hands generally contact the material being sampled. Other protective equipment may also require decontamination. Remember, the most effective procedure is avoidance of contamination (e.g. work neatly). Also, the use of disposable equipment is an alternative to decon. Evaluation of the effectiveness of a decon program is completed using equipment blanks disclosed in a later section.

Table 1. Generalized decontamination solutions used.

Contamination	Typical Cleaning Protocol
Organics	Wash: soapy water (lab-grade detergent) Rinse: solvent (acetone, hexane, isopropanol, or methanol) Rinse: distilled and/or deionized water Air dry
Metals	Wash: soapy water (lab-grade detergent) Rinse: dilute acid solution (10% HCl or HNO_3) Rinse: distilled or deionized water Air dry
Petroleum hydrocarbons	same as organics <u>or</u> Wash: degreasing detergent solution Rinse: distilled &/or deionized water Air dry

1.13 FIELD NOTEBOOK

All observations made during sampling activities should be recorded in a bound field notebook. Not only will the observations in your field book help with data validation and interpretation, this documentation may be entered as evidence should litigation ensue. Following are points that should be followed when using a field notebook.

- Number all pages; retain the first two pages for the table of contents.

- Write in water-insoluble ink. If an error is made, draw a single line through the error, initial and date the error, and write in the correct entry.

- Title the entry and record date, time, weather, name of all crew members, and a brief description of the purpose of the investigation. A rough map of the area is often useful.

- Record all pertinent observations.

- If instruments are used during the investigation, background values followed by site values are recorded. Also record any calibrations and a description of the instrument (make, model, etc).

- Sign your name at the bottom of each page.

- If a page is left blank, write "Page intentionally left blank" across the page and draw a line through the page.

- Write legibly; someone other than yourself may have to read your entries. Avoid abbreviations that might be misinterpreted.

- Retain the field notebook as a permanent record.

1.14 INTERFACING WITH THE LABORATORY

The analytical laboratory is an integral part of every sampling program. Laboratory personnel should be involved prior to, during, and after sampling. All to often, the laboratory is forgotten until they receive the sample and cannot complete the required analysis because of some mistake or misunderstanding on the part of the sampler. Following is a brief review of the role of the laboratory in a site assessment.

1.14.1 Selection of a Laboratory

Once project needs have been established, quality and service should be the most important criteria when selecting a laboratory. Other factors may include:

- project experience;
- experience of personnel;
- certifications, and other approvals;
- written quality assurance/quality control plan;
- use of the lab as a consultant;
- price.

Other procedures that are often useful in selecting a laboratory include references, range of services, special services, accreditations, client list, reputation, turn-around times, data handling, etc. Before the final selection is made, visit the lab and interview key people.

1.14.2 Containers

The laboratory usually supplies the sampling containers. Items such as material, size and number of containers, labeling, preservation, coolers, cooling material, packing material, chain-of-custody forms, labels, etc., should be determined in cooperation with the laboratory. In most cases, the laboratory can supply these items.

1.14.3 Methods of Analysis

Analytical methods depend on what is under investigation, why the investigation is taking place, and the regulatory agency. In many cases the EPA has specified methods of analysis for projects under it's jurisdiction. However, methods change, so both the sampler and the laboratory must be aware of changes to ensure that the proper methods are used. It is essential to clarify analytical methods before sample collection due to the dependency of most methods on sample volume, preservation, holding time, etc.

Detection limits must be set and are based on the needs of the project regulatory requirements, and capabilities of the analytical equipment.

If the analytical program is phased, description of each phase, criteria limits, and other details must be coordinated with the sampler and the laboratory.

1.14.4 Analytical Quality

Methods of laboratory quality assurance and quality control should be examined to ensure that project objectives and standards can be met. Details of this program for both the field and laboratory are discussed in the next section.

1.14.5 Dates

As previously discussed, sample holding times (measured in calendar days) are very important, especially when phased (triggered) analyses are conducted. Coordination between sampler and laboratory is essential to prevent analytical voids due to expired holding times.

Turn-around time (measured in working days) is also important and is probably the most common source of misunderstanding between the sampler and the laboratory. Turn-around time must be specified prior to initiation of analytical work. Most laboratories can "rush" the samples. Depending on the situation, turn-around time can be reduced to a few hours. However, analytical costs are often doubled.

Delivery dates and times must likewise be specified. Someone must be at the laboratory to accept sample custody and to unpack and log in the samples as soon as they arrive. After hours or weekend sample deliveries must be coordinated with the laboratory. Avoid sample deliveries on Thursdays and Fridays because it will likely be the next week before the laboratory addresses the samples.

1.14.6 Laboratory Safety

Samplers are obligated to specify expected concentrations in sample parameters. This allows the laboratory to properly adjust the analytical procedures, thus avoiding costly delays and inaccurate results. In addition, highly concentrated samples can pose a hazard to laboratory personnel and can contaminate the laboratory. This is especially important when samples of pure product, such as pesticides, are sent in for analysis.

With exception of samples that must be collected without any headspace (VOAs), do not overfill sample containers. The sample container should be filled to the lower portion of the neck or to within one inch of the top. This allows laboratory personnel to safely open the container without spilling any of the contents.

Disposal of excess sample material must be addressed. Most laboratories are small quantity hazardous waste generators and operate under RCRA regulations. Laboratories can drum samples and ship them to an incinerator, or send them back to the sampler for proper disposal. The fate of submitted samples should be discussed prior to sample delivery.

1.15 QUALITY ASSURANCE/QUALITY CONTROL

Quality assurance and quality control (QA/QC) is a management program to ensure that the specified standards of quality are maintained throughout the sampling and analytical program. Actually, quality assurance refers to the overall system of activities whose purpose is to provide assurance that the system is in control (the QC is effectively implemented). Quality control includes the actual day-to-day activities to control and assess data quality. The QA/QC plan usually contains such items as project description, sampling procedures, sample custody, equipment calibration, analytical procedures, data reduction, quality control checks, performance audits, precision, accuracy, control samples, and reporting. The degree of quality required for a particular project must be specified within the project DQOs. Some common sense needs to be considered in development of these DQOs. For example, requiring the analytical precision to be less than 1% when the sampling precision is always above 15% is unjustified.

1.15.1 Accuracy and Precision

Accuracy is the degree of agreement of a measured (analytical) value with the true value. Accuracy is extremely difficult to evaluate with environmental and hazardous matrices due to the inability to asses the "true" analytical value. In contrast, precision measures the ability to replicate results. Both accuracy and precision must be evaluated as a part of the QA/QC program.

1.15.2 Quality Control Samples

A variety of quality control samples may be collected in the field to assist QA/QC evaluation both in the field and in the laboratory. These samples may include the following.

- **Field Duplicate, Replicate, or Split**—To measure analytical precision, duplicate samples are taken of the same media, being careful that one sample is the twin of the other. With respect to the laboratory, field duplicates are usually "blind;" in other words, the laboratory does not know the sample is a duplicate. In some cases, sample splits must be offered to the facility owner. The owner may have the samples analyzed by another lab, thus determining analytical precision between labs. For some analyses (such as volatile organics and trihalomethanes), duplicates are required. If the collected sample is non-homogeneous (such as soil), then the sample must be homogenized (except for VOA samples) and split in a representative manner.

- **Field Spike**—a duplicate sample is collected, but to one sample a known amount of a chemical (analyte) is added. This is used to determine analytical accuracy. However, an exact amount of a particular analyte must be added to the sample and this is difficult to accomplish in the field. Additionally, this does not fully take into account "matrix," or the mass that encloses the sample. Field spikes are rarely used.

- **Field Method Blank**—following routine equipment decontamination, the process is repeated and decontamination solution collected for analysis. A sample of unused decontamination solution is also collected. A comparison of the analytical results will determine the effectiveness of field decontamination.

- **Field Blank**—A sampling container is filled with distilled water at the same time a sample is taken. This is commonly used if air-borne contaminants are of concern. This is sometimes called an ambient blank.

- **Field Preservation Blank**—A sampling container is filled with distilled water and a preservative added. This will determine if any contamination is introduced by the preservative.

- **Rinsate Blank**—Distilled or "laboratory grade" water is flushed through a sampling instrument and the rinseate is then sent to the lab for analysis. This type of blank can be collected prior to sampling or following sampling and decontamination. Collecting a rinseate sample is an effective means of demonstrating contamination-free sampling equipment. This is also referred to as an "equipment blank."

- **Trip Blank**—The laboratory fills a container with distilled water and sends it with the other sampling containers. This quality control sample is not opened in the field, but is otherwise treated the same as other samples. Analysis of this sample will determine if any contamination takes place from the time containers leave the laboratory until they return. Contamination may occur when containers are prepared in the lab if samples are not properly capped, improperly cleaned cooler, in appropriate packing material, etc. Trip blanks are required for certain analyses such as volatile organics.

A variety of other quality control samples may be collected depending on DQOs, regulations, the client, etc. For example, co-located samples may be required. These are two samples collected adjacent to one another (such as two soil samples taken within inches of each other). These samples should be equally representative of the parameter of interest. In this respect, co-located samples are similar to a duplicate sample except that co-located samples will reveal any localized spatial variability. If analytical results show significant differences between a set of co-located samples, then more variation than anticipated exists within the system and a different sampling strategy may be necessary.

Background samples are often included in the field quality control program. A background sample is collected in an area not thought to be contaminated by the parameters under investigation. Background samples often have detectable levels of parameters, but their source is not related to the release under investigation.

Material blanks may be collected during some sampling programs. These are samples of construction materials, such as the sand or grout used when installing a ground water well, that may introduce contaminants.

There is considerable debate regarding the nomenclature and types of quality control samples that are taken in the field. Therefore, for a given sampling effort, the above definitions may change slightly.

The frequency of quality control sample collection must be clearly communicated and documented prior to sampling activities. Types and frequency of quality control sample collection are generally specified by the applicable regulatory agency.

In addition to these field control samples, the laboratory has internal quality controls. These controls usually consist of a method blank (distilled water or solvent), and blank spikes (a method blank plus a known amount of analyte) and matrix spikes for determination of accuracy; and matrix spike duplicates and sample duplicates to determine precision. Also, calibration standards are used to help insure instrumental accuracy and precision.

1.16 SAFETY PLAN FOR SAMPLING

Safety issues are important in sampling because contact with hazardous substances is likely. Therefore, prior to sampling, health and safety issues must be considered and a site safety plan prepared. Additionally, depending upon the degree of anticipated hazards, the sampling crew must have appropriate training. OSHA requires a minimum of 40 hours of appropriate health and safety training along with 24 hours of on-the-job training for on-going sampling activities at hazardous waste sites (see 29 CFR 1910.120). Appendix B presents an example of a sampling health and safety plan; following is a brief discussion of major components of such plans.

1.16.1 Purpose

The purpose of the safety plan is to establish personnel protection standards and mandatory safety practices and procedures. The plan should also provide for contingencies that may arise during sampling. The provision in the safety plan should be mandatory for all on-site personnel. All sampling and support personnel should be familiar with the plan and comply with its requirements.

1.16.2 Hazard and Risk Identification

All suspected and confirmed chemical hazards should be identified in terms of where they occur, exposure limits, warning concentrations, and health hazards posed during sampling and accidental release. Physical hazards must also be identified and mitigated to the extent possible prior to sampling. Physical hazards that remain after mitigation must be described in the safety plan.

1.16.3 Personnel Protection

Based on the hazards and risk of exposure, appropriate protective equipment is specified in the safety plan. Skin and eye protection are needed for virtually all sampling programs. Additional protective equipment may be needed, such as an air-purifying respirator (APR) or a self-contained breathing apparatus (SCBA). Also included in personnel protection is site control, air monitoring, medical monitoring, and related safety items.

1.16.4 Decontamination

Decontamination procedures for personnel, sampling equipment, and samples are required. The procedures should include decontamination solutions, location, personnel involved, equipment, proper disposal of waste products, etc. In most cases, contaminated material should be placed in drums, labeled, and left on-site for disposal by the owner.

1.16.5 Contingency Plan

The purpose of the contingency plan is to define emergency response actions. Chain of command during emergencies is essential and must include a list of individuals and their assigned roles. Off-site support must be specified, such as paramedic units and the receiving hospital. Procedures to check sampling personnel for heat stress, cold stress, and chemical exposure must be specified.

1.17 SOURCES OF INFORMATION

Addiscott, T.M., and J.R. Wagenet. 1985. A Simple Method for Combining Soil Properties that Show Variability. Soil Science Society of America Journal. 49: 1365-1369.

Box, G.E.P., W. Hunter, and J.S. Hunter. 1978. Statistics for the Experimenter: An Introduction to Design, Data Analysis, and Model Building. John Wiley and Sons, New York.

Camp Dresser & McKee Inc. (CDM). 1986. Statistics for Contaminated Zones at the North Cavalcade Site, CDM Internal Correspondence, J. Sullivan.

EPA. 1984. A Soil Sampling Quality Assurance User's Guide. EPA 600/4-85-043.

EPA. 1985. Sediment Sampling Quality Assurance User's Guide. EPA 600/4-85-048.

Flatman, G.T. 1985. Design of a Soil Sampling Program: Statistical Considerations Draft.

Flatman, G.T. and A.A. Yfantis. 1984. Geostatistical Strategy for Soil Sampling: The Survey and the Census. Environmental Monitoring and Assessment. 4:335-349.

Gordon, N. D., T. A. McMahon, and B. L. Finlayson. 1992. Steam Hydrology. John Wiley & Sons, New York.

Isaaks, E. 1984. Risk Qualified Mappings for Hazardous Waste Sites: A Case Study in Distribution Free Geostatistics, unpublished master's thesis, Stanford University.

Journel, A.G. 1983. Non Parametric Estimation of Spatial Distribution. Journal of Mathematical Geology. Vol. 15, No. 3, pp. 445-468.

Journel, A.G. and C.J. Huijbregts. 1978. Mining Geostatistics. Academic Press, London.

Keith, L.H. 1991. Environmental Sampling and Analysis: A Practical Guide. Lewis Publishers, Boca Raton, Florida.

Klusman, R.W. 1985. Sample Design and Analysis for Regional Geochemical Studies. Journal of Environmental Quality. 14:369-375

Lambeth, S. 1991. RCRA/CERCLA Cleanups: Considerations for Selecting a Lab. Pollution Equipment News. February 1991.

Ripley, B. 1982. Spatial Statistics. John Wiley & Sons. New York.

Russo, D. 1984. Design of an Optimal Sampling Network for Estimating the Variogram. Soil Science Society of America Journal. 48 (4): 708-716

Sullivan. J. 1984. Conditional Recovery Estimation Through Probability Kriging - Theory and Practice in Geostatistics for Natural Resource Characterization. Reidel. Dordrecht, Holland.

Verly, G. 1983. The Multigaussian Approach and its Application to the Estimation of Local Reserves. Journal of Mathematical Geology, Vol. 15, No. 2, pp. 263-290.

Yost, R.S.G. Uehara and R.L. Fox. 1982. Geostatistical Analysis of Soil Chemical Properties of Large Land Areas. Semi-variograms. Soil Science Society of America Journal 46(5): 1028-1032.

NOTES

APPENDIX A

Following are excerpts from the Field Sampling Plan portion of the Phase I RFI/RI Work Plan for the Woman Creek Drainage at Rocky Flats, Golden, Colorado. This document was dated February 1992.

FIELD SAMPLING PLAN

A1.1 Background and Sampling Rationale

A1.1.1 <u>Background</u>

The objectives of the Phase I RCRA Facility Investigation (RFI)/Remedial Investigation (RI) are:

- To characterize the physical and hydrogeologic setting of the Individual Hazardous Substance Sites (IHSSs).

- To assess the presence or absence of contamination at each site.

- To characterize the nature and extent of contamination at the sites, if present.

- To support the Phase I Baseline Risk Assessment (BRA) and Environmental Evaluation.

Within these broad objectives, site-specific data needs have been identified in another section. The purpose of this section of the work plan is to provide a Field Sampling Plan (FSP) that will address data needs and data quality objectives.

In the Phase I investigation for the Original Landfill (IHSS 115), data will be collected to define contamination boundaries and investigate the potential for contaminant migration. Based on the Phase I investigation results, a Phase II source characterization investigation will be performed. If warranted, an Interim Measures/Interim Remedial Action may also be performed at IHSS 115 once Phase I results are evaluated. Additional phases of investigation and risk assessment may be required at other IHSSs pending the Phase I results, although they are not anticipated at this time.

Generally, only limited information is available concerning the IHSSs in (operable unit number 5) OU 5 since there have been no previous field investigations of these sites. Available information includes aerial photographs, site histories, and some analytical data for samples collected near the IHSSs. Little information exists specific to the physical characteristics of the sites or to the nature and extent of the contamination, if present.

One of the objectives of the RFI/RI is to assess the presence or absence of contamination in the groundwater, surface water, sediments, and soils at the sites. A multi-staged approach is outlined in the IAG and will be used in Phase I to achieve this objective. This technique uses an "Observational Approach" involving continuing reassessment of the site conditions as data are obtained. As data are collected and interpreted, specific sampling plans will be formulated to build on existing information.

A1.1.2 Sampling Rationale

As discussed above, a staged approach will be used for the sampling program. There are four stages that may be completed at any site.

- **Stage 1** consists of a review of existing data, including aerial photographs and site records. Data from ongoing or other OU investigations that have become available since preparation of this Phase I work plan will be compiled and evaluated. These data will be validated as appropriate for incorporation into the OU 5 site characterization. This review of existing information has already been partially performed during preparation of this Phase I work plan.

- **Stage 2** involves screening activities, including radiation, magnetometer, electromagnetic (EM), and soil gas surveys. These activities are designed to provide Phase I screening-level data concerning the presence or absence of contaminants at some of the sites. These surveys will be conducted in the order listed. Each screening activity will be performed after review of the previous screening method.

- **Stage 3** consists of Phase I sampling activities for soil, sediment, and surface water. Soil borings will be completed at some IHSSs to collect samples at depth and to characterize the IHSS. Some of the sampling locations may be selected to investigate anomalies identified in the Stage 2 screening surveys. This stage will provide confirmation of the Phase I screening data as well as aid in Phase I geologic and hydrogeologic characterization of the sites.

- **Stage 4** involves cone penetrometer surveys, monitoring well installation, and groundwater sampling. Cone penetrometers will be used to characterize subsurface lithology, to help locate vadose zone water or groundwater, and to help guide installation of monitoring wells. If pore pressure in the vadose zone indicates the presence of water, a BAT sampler (or equivalent) will be inserted to take a sample. Groundwater monitoring wells will be installed to characterize the hydrogeologic setting of each site and to monitor alluvial groundwater conditions within or downgradient of several sites. These wells will be sampled after completion and development, and the results will be included in the Phase I RFI/RI Report.

- **Stage 5** consists of additional sampling or surveying activities unique to each IHSS.

A1.2 Phase I Investigation Program

This section describes the Phase I investigation program for the IHSSs within OU 5. For each IHSS, the tasks listed are generally divided into office activities prior to field sampling (Stage 1), field screening activities prior to sampling (Stage 2), field sampling activities (Stage 3), and groundwater monitoring, well installation, and sampling (Stage 4). As part of the field sampling program, data from site-wide monitoring programs and investigations at other OUs will be used as appropriate to add to, or substitute for, the data collected during the Phase I investigation.

The sites included within OU 5 are IHSS 115 - Original Landfill; IHSS 133 - Ash Pits 1 - 4, the Incinerator, and the Concrete Wash Pad; IHSS 142.10 and 142.11-C-Series Detention Ponds, and IHSS 209 - Surface Disturbance southeast of Building 881 and two additional surface disturbances; these are the surfaces west of IHSS 209 and the surface disturbances south of the Ash Pits. The area south of OU 5 to the property boundary will be investigated, if warranted. For reference, the Phase I investigation programs for each IHSS are summarized below. A number of SOPs will be used during the investigation.

A1.2.1 IHSS 115 - Original Landfill

Stage 1 - Review Aerial Photographs and Gamma Radiation Survey Results

Aerial photographs taken during operation of the Original Landfill will be reviewed to identify the extent of the Original Landfill and the disturbed area located to the east of the Original Landfill. The areas to be studied during later steps of this investigation, including the location of former pond, will be delineated from the aerial photographs and surveyed on the ground as needed to define their locations for the Phase I field work. Additional studies conducted at the Landfill after preparation of this Phase I work plan will be evaluated during Stage 1. Also as part of this stage, the gamma radiation survey conducted at the Original Landfill in Fall 1990, using a germanium detector will be further reviewed, and the elevated radiation readings shown on Figure 1 will be surveyed on the ground to define their locations.

Stage 2 - Magnetometer, EM, and Soil Gas Surveys

A magnetometer survey will be performed over and downgradient of the Old Landfill and the disturbed area to the east (Figure 1). This survey will be conducted on a 25-foot grid in the area outlined for the radiation survey in Figure 1. The survey will be completed according to the magnetic locator procedure in SOP GT. 10. Resulting anomalies will be mapped and contoured.

An EM geophysical survey will be performed over the Old Landfill on the same 25-foot grid established for the magnetometer survey and will cover the same area. The survey will be completed according to the EM geophysical procedures in SOP GT. 18. Details of both the magnetometer and EM geophysical survey will be supplied to the Agencies for review in a TM. The TM will include the type of geophysical surveys to be performed, procedures, and grid spacing.

A real-time soil gas survey will be conducted over the Original Landfill and the disturbed area located to the east of the Landfill (Figure 1) to identify areas of volatile organic contamination. As specified in the IAG, the soil gas sample will be taken on a 100-foot grid according to the procedures described in SOP GT. 9. To further improve the sampling coverage, the grid will be reduced to 25-foot spacing at the downgradient perimeter of the landfill, over areas of suspected buried metallic materials based on the magnetometer and EM survey, and over areas where volatiles are found during the 100-foot grid soil gas survey. The perimeter of the landfill will be defined by the aerial photograph interpretation, radiation, magnetometer, and EM survey review, and by field reconnaissance. The 25-foot soil gas grid spacing around the downgradient perimeter will cover at least the area between the lat 100-foot grid location within the landfill area and the first 100-foot grid location outside the landfill area (see Figure 1). The 25-foot soil gas grid located over metallic materials or volatile plumes will continue for at least 50 feet

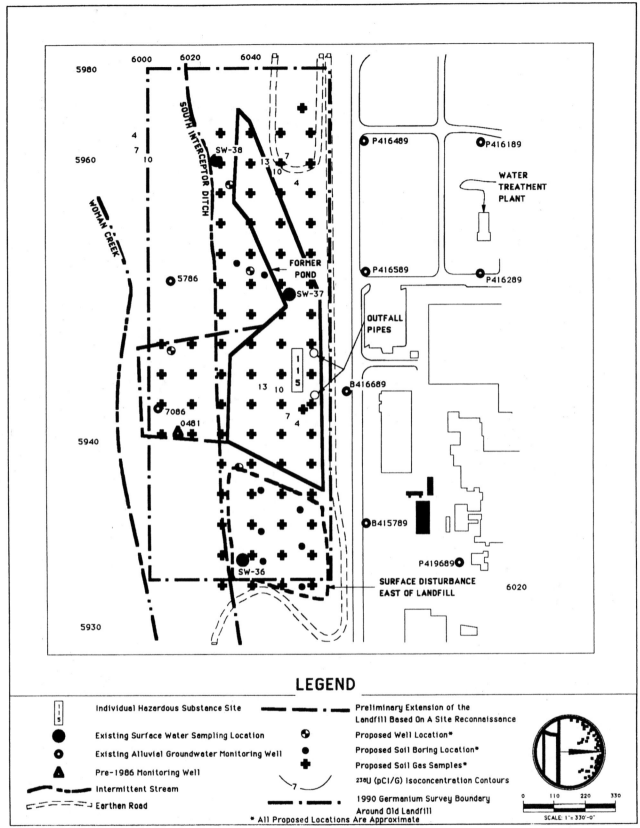

Figure 1 Sampling and well locations.

beyond the edge of the anomaly. This approach should better characterize the area of likely contamination. A probe will be driven approximately 5 feet into the soil to collect the soil gas. The soil gas sample will be analyzed for 1, 1, 1-trichloroethane (TCA), dichloromethane, benzene, carbon tetrachloride, tetrachloroethylene (PCE), and trichloroethylene (TCE) using a portable gas chromatograph (GC). Analytical peaks of compounds for which the GC is not calibrated will be noted. It will not be possible to analyze for solvent breakdown products like 1,2-dichloroethane and vinyl chloride with a GC because they co-elute with other compounds. Vinyl chloride co-elutes with freon compounds, and 1,2-dichloroethane co-elutes with methyl ethyl ketone and dibromomethane.

Stage 3 - Surface Soil, Soil Core, Soil Boring, Sediment, and Surface Water Samples

Randomly located surficial soil samples will be collected to characterize the landfill cover material and exposed fill material using the Rocky Flats method. Depending upon the results of the radiation screening, additional surface soil samples may be required at identified areas with above background radiation. These samples will help establish whether the landfill is leaking via fugitive dust entrained in air for risk assessment purposes. In addition, based on the review of the gamma radiation survey, additional surficial soil sample will be collected within thee areas that have above background radiation. At least two samples will be collected at small or point sources of radiation and at least three will be collected over disturbed areas. A TM will be submitted to the agencies for review prior to implementation that will specify the exact number of samples necessary for the risk assessment, and identify the sampling locations and sampling method protocol.

Soil cores will be collected on a random basis to verify the soil gas survey and other screening methods (e.g. false negative). One soil core (grab sample) will be collected for every 15 to 20 soil gas samples at the same depth as the soil gas samples. Based on the number of original grid soil gas sampling locations, it is estimated that four soil cores will be collected.

Three soil borings will be placed at up to three areas where plumes have been identified by the soil gas survey. This will result in a maximum of nine soil borings being drilled at the three plume areas. At each plume area, one soil boring will be placed at the point of the highest soil gas reading, and two borings will be located downslope of that point within the plume identified by the soil gas survey.

Soil borings will also be drilled for subsurface characterization purposes. One soil boring will be drilled in the location of each of the two former ponds. Six soil borings will be drilled in the disturbed area east of the landfill. Each soil boring will be drilled at least 6 feet below the base of the alluvial material according to the procedures described in SOP GT.2. Samples will be taken continuously in these borings. Discrete samples will be collected from every 2-foot increment and analyzed for the TCL volatile organic compounds (VOCs). Samples will be composited from every 6-foot interval and analyzed for the TCL semivolatile organic compounds, the TAL metals, and radionuclides. As specified in the SOP, samples will not be collected for chemical analysis from the saturated alluvium.

During sampling a soil classification survey will be completed at the Original Landfill for use in the Environmental Evaluation. Several samples may also be collected from 0 to 2 feet for grain size analysis.

The sediments and surface water of the South Interceptor Ditch (SID) and Woman Creek will be sampled immediately downgradient of the Original Landfill. Surface water samples will be collected at three locations along the SID and three locations on Woman Creek (total of six samples) according to the procedures specified in SOPs SW.2 and SW.3 for surface water. Sediment samples will be collected at two locations along the SID and two locations on Woman Creek (total of four samples) according to procedures specified in SOP SW.6. The sediment samples will be collected in areas of the creek or ditch that are conducive to sediment accumulation.

Stage 4 - Cone Penetrometer, BAT Sampler (or equivalent), Monitoring Well Installation and Groundwater Sampling

A cone penetrometer will be used to establish subsurface conditions and lithologies downgradient from the landfill. One subsurface condition that is essential to characterize is soil moisture and/or saturation. A cone penetrometer with this capability will be used. Two lines of cone penetrometer surveys will be taken with a maximum of 100-foot spacing between penetrometers; one line will be between the Landfill and the SID, and one line between the SID and Woman Creek (see Figure 1). In the appropriate cone penetrometer survey locations (locations where significant soil moisture is present), BAT sampling, or an equivalent, will be used to sample any encountered groundwater or interstitial fluid. These samples are necessary to help establish whether contaminated plumes are presently leaking from the landfill. To calibrate the cone penetrometer, one of the soil borings discussed above will be "twinned" so that the cone penetrometer will penetrate known lithologies and saturations. A TM will be submitted to the Agencies for review outlining the details of the cone penetrometer use, type of sampler, spacing and analyte list.

Base on information from the magnometer, EM, soil gas surveys, and cone penetrometer data, the location for alluvial monitoring wells will be determined. Final locations for the monitor wells will be submitted to the Agencies for review in a TM. It is possible due to the limited saturated thickness of the alluvium, that there may be locations where there is no water or times of the year when the saturated thickness is zero. If this is the case, it may be necessary to relocate the wells or possibly install a vadose zone sampling device such as the BAT sampler (or an equivalent) capable of characterizing the contaminant plumes in zones of limited water. It may also be necessary to install bedrock wells beneath zones of contaminated alluvial ground water or if a subcropping sandstone is encountered. The need for bedrock wells will be evaluated after lithologic and preliminary chemistry data has been gathered and interpreted. At this time it is proposed that a maximum of three monitoring wells will be installed in these borings. All of these wells will be installed in the alluvium just above the bedrock according to SOP GT.6.

In addition to the above wells, four alluvial monitoring wells and/or vadose zone samples will be installed in the alluvium downgradient of the Original Landfill. The location, type, and number of monitoring devices will be dependent upon the result of all other data gathered in this Phase I investigation. At this time, it appears at least three wells should be installed between the Landfill and the SID, and one well installed between the SID and Woman Creek: locations shown on Figure 1 are tentative. The first well will be placed approximately between the western leg of the landfill and the SID. The second well will be place in the alluvium in the surface drainage north of Well 5786 between the Landfill and the SID within the area of the old embankment. The third well will be placed in the alluvium between the southeastern corner of the boundary of IHSS 115 and the SID, downgradient of the outfall identified on the east side of the landfill. The fourth well will be placed between existing wells 5786 and 7086. These locations may be modified slightly depending upon the results of the screening surveys. If a water-bearing sandstone unit is found to be the first bedrock unit underlying the alluvium in a boring, then an additional well will be completed in the sandstone at that location. The use and location of the proper type of monitoring device should be able to ascertain both present and future contaminant levels and help establish any future or present contaminant migration problems. The locations for the monitoring devices should allow for monitoring the principal groundwater and downgradient migration pathways of the Old Landfill.

All groundwater monitoring wells will be drilled according to SOP GT.2 and installed according to SOP GT.6. All wells will be developed according to SOP GW.2. Following development, wells will be sampled according to SOPs GW.5 and GW.6. The results of the first round of sampling will be reported in the Phase I RI Report. The four monitoring wells downgradient of the Landfill will be sampled quarterly for a minimum 1 year.

Stage 5 - Outfall Pipe Location, Source, and Sampling

The two corrugated metal pipes protruding from the Landfill (Figure 1) will also be investigated in this FSP. Plant plans will be reviewed and a sewer snake survey will be conducted to attempt to identify the open length of the pipes and the sources of water. This survey may use a traceable electronic or magnetic source attached to the snake such that surface instruments can be used to follow the path of the pipe. Other methods for locating pipes may also be used if the sewer snake survey is inconclusive. If water is found to be flowing through either of the corrugated pipes during this Phase I investigation, the effluent will be sampled according to SOP SW.3. Results of the sampling will be reported in the Phase I RI Report.

A1.3 Sample Analysis

This section describes the sample handling procedures and analytical program for samples collected from the Phase I investigation. In this section, sample designations, analytical requirements, sample containers and preservation, and sample handling and documentation requirements will be discussed.

A1.3.1 Sample Designations

All sample designations generated for this RFI/RI will conform to the input requirements of the Rocky Flats Environmental Database System (RFEDS). Each sample designation will contain a nine-character sample number consisting of a two-letter prefix identifying the media sampled (e.g., "SB: for soil borings, "SS: for stream sediments), a unique five-digit number, and a two-letter suffix identifying the contractor (e.g., "WC" for Woodward-Clyde). One sample number will be required for each sample generated, including QA/QC samples. In this manner, 99,999 unique sample numbers are available for each contractor that contributes sample data to the database. A block of numbers will be reserved for the Phase I RFI/RI sampling of OU 5. Boring numbers will be developed independently of the sample numbers from a boring. Specific sample location numbers are not assigned at this time, pending the results of the aerial photograph analysis and review of existing data.

A1.3.2 Analytical Requirements

General, samples collected during the Phase I RI will be analyzed for some or all of the following chemical and radionuclide parameters:

- Nitrates
- TAL metals
- Uranium 233, 234, 235, and 238
- Transuranic elements (plutonium and americium)
- Cesium 137 and strontium 89, 90
- Gross alpha and gross beta
- Tritium
- Total dissolved chromium (water only)
- Beryllium
- TCL volatile organics
- TCL semivolatile organics
- Total organic carbon (TOC)
- TCL pesticides/PCBs
- CO_3, HCO_3, Cl, SO_4, NO_3 (water only)

For the specific analytes in the group listed above, their detection/quantitation limits and analytical methods are presented in the appendix. Both filtered and unfiltered surface water and groundwater samples will be analyzed at each location. The analytical program for each IHSS was developed in the IAG based on the type of waste suspected to be present at each site.

Nitrates are included because low-level radioactive wastes with high nitrate concentrations may be present in Woman Creek or the SID. Metals were probably disposed of at OU 5; however, details are not well known. Therefore, all of the TAL metals have been selected for Phase I analysis.

Uranium is likely to have been a constituent of the wastes at OU 5. The isotopes U-233, U-234, U-235, and U-238 have been selected for analysis in Phase I. Plutonium is the only transuranic element that is used on the site. However, americium is a daughter product of plutonium and is found at the Rocky Flats Plant. Therefore, plutonium and americium have also been selected as Phase I radionuclide parameters. Gross alpha and gross beta are included as screening parameters because they are useful indicators of radionuclides. Tritium, strontium, and cesium are also included in the analytical program.

Volatile and semivolatile organics may have been handled at OU 5 in small quantities probably only at the Original Landfill. The specific compounds used are unknown; therefore, all of the TCL volatile and semivolatile organics will be included in the Phase I analyses for some samples.

TCL pesticides/PCBs and TOC have been included for some samples to provide data for the environmental evaluation. For the sediment sample collected from Woman Creek and the SID, TCL pesticides will be analyzed in the samples collected from the detention ponds and at the location just downgradient from the Original Landfill. The other sediment samples collected from Woman Creek and the SID will not be analyzed for TCL pesticides as no pesticides have been detected to date from the extensive sampling already performed. In addition, the two proposed sediment sampling locations just downstream of the Ash Pit will not be analyzed for TCL volatiles and semivolatiles since incineration would probably have destroyed these organics.

The analytical parameters for the soil gas survey at IHSS 115 are 1,1,1-trichloroethane (TCA), dichloromethane, benzene, carbon tetrachloride, tetrachloroethylene (PCE), and trichloroethylene (TCE).

A1.3.3 Sample Containers and Preservation

Sample volume requirements, preservation techniques, holding times, and container material requirements are dictated by the media being sampled and by the analyses to be performed. The soil matrices to be analyzed will include soils and sediments. The water matrices for analysis will include surface water and groundwater. Table 1 and 2 list analytical parameters of interest in OU 5 for water and soil matrices, along with the associated container size, preservatives (chemical and/or temperature), and holding times. Additional specific guidance on the appropriate use of containers and preservatives is provided in SOP OF.13, Containerizing, Preserving, Handling, and Shipping of Soil and Water Samples.

A1.3.4 Sample Handling and Documentation

Sample control and documentation is necessary to ensure the defensibility of data and to verify the quality and quantity of work performed in the field. Accountable documents include logbooks, data collection forms, sample labels or tags, chain-of-custody forms, photographs, and analytical records and reports. Specific guidance defining the necessary sample control, identification, and chain-of-custody documentation is discussed in SOP OF. 14.

A1.3.5 Data Reporting Requirements

Field data will be put into the RFEDS using a remote data entry module supplied by EG&G. Data will be entered on a timely basis and a 3.5-inch diskette will be delivered to EG&G. A hard copy report will be generated from the module for contractor use. The data will be put through a prescribed QC process based on SOP OF.14 to be generated by EG&G.

A sample tracking spreadsheet will be maintained by the contractor for use in tracking sample collection and shipment. EG&G will supply the spreadsheet format and will stipulate the timely reporting of the information. This data will also be delivered to EG&G on 3.5-inch diskettes. Computer hardware and software requirements for contractors using government supplied equipment will be supplied by EG&G.

A1.4 Field QC Procedures

Sample duplicates, field preservation blanks, and equipment rinsate blanks will be prepared. Trip blanks will be obtained from the laboratory. The analytical results obtained for these samples will be used by the Environmental Restoration (ER) Project Manager to assess the quality of the field sampling effort. The types of field QC samples to be collected and their applications are discussed below. The frequency for QC samples to be collected and analyzed is provided in Table 3.

Duplicate samples will be collected by the sampling team and will be used as a relative measure of the precision of the sample collection process. These samples will be collected at the same time, using the same procedures, the same equipment, and in the same types of containers as required for the samples. They will also be preserved in the same manner and submitted for the same analyses as required for the samples.

Field preservation blanks of distilled water, preserved according to the preservation requirements, will be prepared by the sampling team and will be used to provide an indication of any contamination introduced during the field sample preparation technique. These QC samples are applicable only to samples requiring chemical preservation. Equipment (rinsate) blanks will be collected from a final decontamination rinse to evaluate the success of the field sampling team's decontamination efforts on nondedicated sampling equipment.

Equipment blanks are obtained by rinsing cleaned equipment with distilled water prior to sample collection. The rinsate is collected and placed in the appropriate sample container. Equipment rinsate blanks are applicable to all analyses for water and soil samples.

Trip blanks consisting of deionized water will be prepared by the laboratory technician and will accompany each shipment of water samples for volatile organic analysis. Trip blanks will be stored with the group of samples with which they are associated. Analysis of the trip blank will indicate migration of volatile organics or problems associated with the shipment, handling, or storage of the samples.

TABLE 1

SAMPLE CONTAINERS, SAMPLE PRESERVATION, AND SAMPLE HOLDING TIMES FOR WATER SAMPLES

Parameter	Container	Preservative	Holding Time
Liquid - Low to Medium Concentration Samples			
Organic Compounds:			
Purgeable Organics (VOCs)	2 x 40-ml VOA vials with teflon-lined septum lids	Cool, 4°C[a] with HCl to pH<2	7 days
Extractable Organics (BNAs), Pesticides and PCBs	1x 4-l amber[b] glass bottle	Cool, 4°C	7 days until extraction, 40 days after
Inorganic Compounds:			
Metals (TAL)	1 x 1-l polyethylene bottle	Nitric acid pH<2; Cool, 4°C	180 days[c]
Cyanide	1 x 1-l polyethylene bottle	Sodium hydroxide[d] pH>12; Cool, 4°C	14 days
Anions	1 x 1-l polyethylene bottle	Cool, 4°C	14 days
Sulfide	1 x 1-l polyethylene bottle	1 ml-zinc acetate sodium hydroxide to pH>9; Cool, 4°C	7 days
Nitrate	1 x 1-l polyethylene bottle	Cool, 4°C	48 hours
Total Dissolved Solid (TDS)	1 x 1-l polyethylene bottle	Cool, 4°C	48 hours
Radionuclides	1 x 1-l polyethylene bottle	Nitric acid pH <2;	180 days

[a] Add 0.008% sodium thiosulfate ($Na_2S_2O_3$) in the presence of residual chlorine.

[b] Container requirement is for any or all of the parameters given.

[c] Holding time for mercury is 28 days.

[d] Use ascorbic acid only if the sample contains residual chlorine. Test a drop of sample with potassium iodine-starch test paper; a blue color indicates need for treatment. Add ascorbic acid, a few crystals at a time, until a drop of sample produces no color on the indicator paper. Then add an additional 0.6g of ascorbic acid for each liter of sample volume.

TABLE 2

SAMPLE CONTAINERS, SAMPLE PRESERVATION, AND SAMPLE HOLDING TIMES FOR SOIL SAMPLES

Parameter	Container	Preservative	Holding Time
<u>Soil or Sediment Samples - Low to Medium Concentration</u>			
Organic Compounds:			
Purgeable Organics (VOCs)	1 x 4-oz wide mouth teflon-lined glass vials	Cool, 4°C	7 days
Extractable Organics (BNAs), Pesticides and PCBs	1 x 8-oz wide-mouth teflon-lined glass vials	Cool, 4°C	7 days until extraction, 40 days after extraction
Inorganic Compounds:			
Metals (TAL)	1 x 8-oz wide-mouth glass jar	Cool, 4°C	180 days[1]
Cyanide	1 x 8-oz wide-mouth glass jar	Cool, 4°C	28 days
Sulfide	1 x 8-oz wide-mouth glass jar	Cool, 4°C	14 days
Nitrate	1 x 8-oz wide-mouth glass jar	Cool, 4°C	28 days
Radionuclides	1 x 1-l wide-mouth glass jar	None	45 days

---

[1] Holding time for mercury is 28 days.

TABLE 3

FIELD QC SAMPLE FREQUENCY

Sample Type	Type of Analysis	Media	
		Solids	Liquids
Duplicates	Organics	1/10	1/10
	Inorganics	1/10	1/10
	Radionuclides	1/10	1/10
Field Preservation Blanks	Organics	NA	NA
	Inorganics	NA	1/20
	Radionuclides	NA	1/20
Equipment Rinsate Blanks	Organics	1/20	1/20
	Inorganics	1/20	1/20
	Radionuclides	1/20	1/20
Trip Blanks	Organics (Volatiles)	NR	1/20
	Inorganics	NR	NR
	Radionuclides	NR	NR

NA = Not Applicable
NR = Not Required

APPENDIX B

SAMPLING HEALTH AND SAFETY PLAN

Following is a Health and Safety Plan for a RCRA sampling program at the Oco Refinery in Texas.

B1.1 Purpose and Policy

The purpose of this plan is to establish personnel protection standards and mandatory safety practices and procedures. The plan also provides for contingencies that may arise during field investigations and operations. Appendix 2 of this plan contains the Contingency Plan and the Emergency Action Plan for the Oco and TRC refineries, respectively.

The provisions of this plan and refinery plans are mandatory for all on-site investigations. All Envirotech personnel shall abide by these plans. All personnel who engage in field investigation activities shall be familiar with these plans and comply with their requirements.

A site description and scope-of-work for the project are provided in Section B1.2 and B1.3 presents the project team organization, personnel responsibilities and lines of authority. Safety and health risk analysis along with medical monitoring requirements are contained in Section B1.4. Section B1.5 contains the site emergency response plan and a list of emergency contacts. Requirements for levels of protection are included in Section B1.6, along with air monitoring procedures.

Site control measures and decontamination procedures are contained in Section B.1.7.

B1.2 Site Description and Scope-of-Work

B1.2.1 Background

Envirotech shall be conducting several tasks at the Oco refinery in Commerce City, Texas. The following section details site specific information and general waste characteristics:

- Site: Oco Refinery
- Site Contact: Mr. Jim Jones 490-5087
- Site Safety Manager: Mr. Bill Lusk 490-5043
- Site: Texas Refining Company
- Site Contact: Mr. Randy Matsus 281-2451
- Site Safety Manager: Mr. Ken Goes 275-2450

Overall Hazard is: Serious __x__   Moderate ____
                    Low   ____   Unknown ____

                    Sludge __x__   Gas   ____

Characteristic(s):   Corrosive   __x__   Ignitable   __x__
                      Radioactive _____   Volatile   __x__
                      Toxic   __x__   Reactive   __x__
                      Unknown   _____   Other (Name) _____

B1.2.2 Potential Hazards

The potential hazards to personnel involved in the field investigation include the following:

- Chemical exposure to benzene, toluene, ethylbenzene, and xylene (BTEX), vinyl chloride, Trans-1, 2-dichloroethene, trichlorethane, trichloroethylene, tetrachloroethane, dichlorobenzene, napthalene, 2-methylnapthaylene, phenanthrene and anthracene.

- Potential for fire or explosion during sampling activities exists. A number of the substances mentioned above as health hazards also create the potential for fire or explosion.

- Potential physical injury during sampling activities.

B1.3   Project Team Organization

The project team assigned to the Oco and TRC Refinery site, their responsibilities, and the lines of authority are outlined below:

Personnel	Task Assigned
J.F. Metlock	Project Manager
J. Hurley	Senior Chemical Engineer
J.A. Krug	Senior Atmospheric Scientist Geostatistician
D. Blane	Project Hydrogeologist
C. Hose	Project Chemical Engineer
R.E. Smith	Project Hydrogeologist; Health and Safety Officer
E. Rowl	Project Environmental Scientist, QA/QC, and Data Base Management
J. Siska	Staff Chemical Engineer
M. Hank	Staff Hydrogeologist
C. Apt	Staff Hydrologist

Assigned project personnel may change over the course of the project. Senior level management for the Oco Refinery and TRC project shall be provided by Mr. Metlock. Mr. Metlock shall be responsible for contractual matters, for allocating resources for the project and shall be responsible for QA/QC and the overall conduct of the project. All field team members are responsible for reading and conforming to the project health and safety plan. No employee shall perform a project activity that he or she believes may endanger his or her health and safety or the health and safety of others.

B1.4 Health and Safety Risk Analysis and Medical monitoring Requirements

Hazardous substances that may be present at the Oco Refinery and TRC include petroleum hydrocarbons, petroleum refinery by-products, and volatile and semi-volatile organic compounds.

Table 1 presents the exposure limits, warning concentrations, and health hazards associated with substances of concern. Many of the hazards substances potentially present on-site are skin and eye irritants. Therefore, skin and eye protection shall be necessary at all times.

In addition to the hazardous substances present on site, some physical or hazardous conditions may be expected at the site. These include risk of injury while working, explosive or combustible atmospheres, and heat stress or exposure to excessive cold.

Employees must implement safe work practice while working on site. Protective clothing shall reduce many of the on-site risks, however, protective clothing and respiratory protection shall increase the potential for heat stress. Heat stress monitoring is discussed in detail in section B2.6.

B1.5 Emergency Response Plan and List of Emergency Contacts

All site activities present a degree of risk to on-site personnel. During routine operations, risk is minimized by establishing good work practices, staying alert, and using proper personal protective equipment. Unexpected events such as physical injury, chemical exposure, or fire may occur and must be anticipated. Envirotech employees are encouraged to participate in Red Cross first aid and CPR courses in order to more effectively handle physical and medical emergencies that may arise in the field.

A high degree of coordination between Envirotech, Oco, and TRC personnel shall be necessary, for both security and safety reasons. Personnel access to the Oco and TRC Refineries is restricted and daily permits must be obtained before field activities begin. The field supervisors shall contact the Oco or TRC representative upon arrival and establish a liaison for security and personnel safety.

B1.5.1 Guidelines for Health and Safety Planning and Training

Employees shall read the site health and safety plan and must familiarize themselves with the information in this chapter. Prior to project initiation, the project or site health and safety officer shall conduct a meeting with the field team members to review the provisions of the health and safety plan and to review the emergency response plan. Employees shall be required to have a copy of the emergency contacts and phone numbers immediately accessible on site and to know the route to the nearest emergency medical services. Employees shall be familiar with each facility's safety policies and shall have completed an OSHA approved 40-hour health and safety training course.

TABLE 1

HEALTH HAZARD QUALITIES OF SUBSTANCES OF CONCERN

COMPOUND	PEL[1] (mg/m^3)	TLV$_{21}$ (mg/m^3)	IDLH[1] ppm	Warning Concentration[4] ppm	Health Hazard
Arsenic	0.5	0.2	-	-	Carcinogen.
Barium	0.5	0.5	250	-	Eye, upper respiratory, skin irritant, slow pulse, spasms.
Benzen	1 ppm	10 ppm	2,000	4.7	Eye, nose irritant; headache, nausea, abdominal pain.
Cadmium (fume)	0.1	0.05	40	-	Nose and throat irritant; cough, chest pain, sweating chills, carcinogen.
Chromium	0.5	0.5	250	-	Eye and skin irritant.
1,4-Dichlorobenzene	450 (75 ppm)	75 ppm	1,000	15-30	Potential carcinogen. Can affect liver.
1,2-Dichlorobenzene	300 (50 ppm)	50 ppm	1,000	-	Eye and nose irritant.
Ethyl Benzene	100 ppm	100 ppm	2,000	0.25-200	Eye and mucous membrane irritant, headache, narcosis, comma.
2-Methylnaphthalene	-	-	-	-	No information available.
Naptha	400(100 ppm)	-	10,000	-	Eye, nose, and skin irritant, drowsiness.
Napthalene	50 (10 ppm)	10 ppm	500	-	Can cause nausea, headache, liver damage.
Toluene	200 ppm	100 ppm	2,000	0.24-400	Dizziness, headache, fatigue, insomnia.
1,1,1-Trichloroethane	1,900	350 ppm	1,000	20-400	Eye and nose irritant, CNS depression.
Trichloroethyloene	100 ppm	270	1,000	-	Eye and skin irritant, vertigo, vomiting, carcinogen.
Vinyl Chloride	1 ppm	10	-	-	Weakness, abdominal pain, carcinogen.
Xylenes	435	100 ppm	10,000	0.05-200	Dizziness, drowsiness, irritant, vomiting, abdominal pain.

B1.5.2 Emergency Recognition and Prevention

Emergency conditions are considered to exist if:

- Any member of the field crew is involved in an accident or experiences any adverse effects or symptoms of exposure while on-site.

- A condition is discovered that suggests the existence of a situation more hazardous than anticipated.

Some ways to prevent emergency situations are listed below:

- Visual contact must be maintained between pairs of field personnel. Team members shall remain close together to assist each other during emergencies.

- During continual operations, on-site workers act as safety backup to each other. Off-site personnel provide support and emergency assistance.

- All field crew members shall make use of their senses (all senses) to alert themselves to potentially dangerous situations which they shall avoid, e.g., presence of strong and irritating or nauseating odors.

- Personnel shall practice unfamiliar operations prior to performing the actual procedure in the field.

- Field crew members shall be familiar with the physical characteristics of investigations, including:

    — Wind direction in relation to contamination zones:
    — Accessibility to associates, equipment and vehicles;
    — Communications;
    — Hot zone (areas of known or suspected contamination);
    — Site access;
    — Nearest water sources; and
    — Emergency contacts.

- Personnel and equipment in the contaminated area shall be minimized, consistent with effective site operations.

- Work areas for various operational activities must be established.

In the event that any member of the field crew experiences any adverse effects or symptoms of exposure while on the site, the entire field crew shall immediately halt work and act according to the instruction provided by the site safety officer.

The discovery of any condition that would suggest the existence of a situation more hazardous than anticipated shall result in the evacuation of the field team and re-evaluation of the hazard and the level of protection required.

In the event an accident occurs, the field supervisor is to complete an Accident Report Form. Follow up action shall be taken to correct the situation that caused the accident. Accident reports shall be submitted to the facility safety manager.

General emergency procedures and specific procedures for handling personal injury and chemical exposure, are described in the following sections.

B1.5.3 Personnel Roles, Lines of Authority and Communication Procedures During Emergency

When an emergency occurs, decisive action is required. Rapidly made choices may have far reaching, long-term consequences. Delays of minutes can create life threatening situations. Personnel must be ready to respond to emergency situations immediately. All personnel shall know their own responsibilities during an emergency, know who is in charge during an emergency and the extent of their authority. This section outlines personnel roles, lines of authority, and communication procedures during emergencies.

In the event of an emergency situation at the site, the field team leader shall assume total control and shall be responsible for all on-site decision making. The designated alternate for the field team leader shall be the site safety officer. These individuals have the authority to resolve all disputes about health and safety requirements and precautions. They shall also be responsible for coordinating all activities until emergency response teams (ambulance, fire department, etc.) arrive on site.

The field team leader and/or site safety officer shall ensure that the necessary Oco, TRC and Envirotech personnel and agencies are contacted immediately after the emergency occurs.

All on-site personnel must know the location of the nearest phone and the location of the emergency phone number list.

B1.5.4 Evacuation Routes and Procedures, Safe Distances and Places of Refuge

In the event of emergency conditions, employees shall evacuate the area, transport injured personnel, or take other measures to correct the situation. An exclusion zone shall be established around the sampling location or other site features and evacuation routes and safe distances decided upon by the field team prior to initiating work. The site safety officer shall be consulted. Generally, the evacuation route shall be placed in the predominantly upwind direction of the exclusion zone.

Consider the mobility constraints of personnel wearing protective equipment. They may have difficulty crossing even small drainage and climbing hills.

B1.5.5 Decontamination of Personnel During an Emergency

Procedures for leaving a contaminated area must be planned and implemented prior to going on site. Work areas and decontamination procedures must be established based on expected site conditions. If a member of the field crew is exposed to chemicals, the emergency procedure outlined below shall be followed:

- Another team member (buddy) shall remove the individual from the immediate area of contamination.

- Precautions shall be taken to avoid exposure of other individuals to the chemical.

- If the chemical is on the individual's clothing, the clothing shall be removed if it is safe to do so.

- Administer first aid and transport the victim to the nearest medical facility, if necessary.

If uninjured employees are required to evacuate a contaminated area in an emergency situation, emergency decontamination procedures shall be followed. At a minimum these would involve moving into a safe area and removing protective equipment. care shall be taken to minimize contamination of the safe area and personnel. Contaminated clothing shall be place in plastic garbage bags or other suitable containers. Employees shall wash or shower as soon as possible.

B1.5.6 Emergency Site Security and Control

The Oco and TRC Refineries are controlled access facilities. Only authorized personnel are allowed to pass the guard checkpoints. Therefore, emergency site security shall not be a problem.

For this project, the field team leader (or designated representative) must know who is on-site and who is in the exclusion zone. Personnel access into the exclusion zone must be controlled. In an emergency situation, only necessary rescue and response personnel shall be allowed into the exclusion zone.

B1.5.7 Procedures for Emergency Medical Treatment and First Aid

Chemical Exposure

In the event of chemical exposure (skin contact, inhalation, ingestion) the following procedures shall be implemented:

- Another team member (buddy) shall remove the individual from the immediate area of contamination.

- Precautions shall be taken to avoid exposure of other individuals to the chemical.

- If the chemical is on the individual's clothing, the clothing shall be removed if it is safe to do so.

- If the chemical has contacted the skin, the skin shall be washed with copious amounts of water, preferably under a shower.

- In case of eye contact, an emergency eye wash shall be used. Eyes shall be washed for at least 15 minutes.

- If necessary, the victim shall be transported to the nearest hospital or medical center. If necessary, an ambulance shall be called to transport the victim. It may be necessary to wrap the victim in a blanket or plastic to avoid contamination of the transport vehicle or hospital emergency room.

Personal Injury

In the event of personal injury:

- Field team members trained in first aid can administer treatment to an injured worker.

- The victim shall be transported to the nearest hospital or medical center. If necessary, an ambulance shall be called to transport the victim.

- The field supervisor is responsible for the completion of an Accident Report Form and the immediate notification of the facility safety manager.

Fire or Explosion

In the event of fire or explosion or if vapor concentrations approach or exceed 25 percent of the Lower explosive Limit (as indicated by a reading on the combustible gas indicator) personnel shall evacuate the area immediately and administer necessary first aid to injured employees. Personnel shall proceed to a safe area and contact the Oco or TRC emergency support services. Upon contacting the emergency support services, state your name, nature of the hazard (fire, high combustible vapor levels), the location of the drill rig or incident, and whether there were any physical injuries requiring an ambulance. Do not hang up until emergency support services has all of the additional information they may require.

Emergency Contact

In the event of any emergency situation or unplanned occurrence requiring assistance, the appropriate contact(s) shall be made from the list below. For emergency situations, telephone or radio contact shall be made with the site point of contact or site emergency personnel who shall then contact the appropriate response teams.

Contingency Contacts	Phone Numbers
Fire Department	218-1535 or 911
Police	218-1535 or 911
Texas Poison Control Center	1-800-312-3073 or 619-1123

<u>Oco</u>
Site Contact: Mr. Jim Jones	490-5087
Site Safety Manager: Mr. Bill Lusk	490-5043
Site Emergency Number (Plant Protection)	490-5050
Main Gate Guard	490-5063

TRC
Site Contact: Mr. Randy Matsus   281-2451
Site Safety Manager: Mr. Ken Goes   275-2450
Ambulance Service Company   889-5151 or 911
Mercy Hospital   793-3000

Envirotech Contacts
Jay Metlock, Project Manager   213-2703
Robert Smith, Project Health and Safety Officer   213-2703

Medical Center

Mercy Medical Center is the closest medical facility to the Oco and TRC refineries. To get there take Brighton Boulevard south to 56th Avenue, take 56th east to Texas Boulevard, take Texas Boulevard to 17th Avenue, head west to Filmore, turn left and proceed to 1650 Filmore.

MERCY HOSPITAL: 793-3000

B1.6   Required Levels of Protection and Air Monitoring

B1.6.1   <u>Personal Protective Equipment</u>

The personal protection level prescribe for most field activities at the Oco and TRC Refineries is Level D with a contingency for the use of Level C or Level B. This requirement is based upon the expected exposure to chemical contaminants and the nature of work to be performed.

Ambient air monitoring of organic gases/vapors, by a Flame Ionization Detector calibrated with benzene standard, shall be used to select the appropriate level of personal protection. The following criteria shall be used:

AMBIENT AIR CONCENTRATION

Organic Gas/Vapor	Protection Level	Respiratory Protection
0 - 1 ppm above background	Level D	None
1 - 5 ppm above background	Level C	APR
5 - 500 ppm above background	Level B	SCBA

All on-site personnel shall have full-face, air purifying respirators equipped with MSA GMC-H cartridges immediately available in the event that Level C respiratory protection is needed. Level B protective equipment shall also be readily available in the event it is warranted. An exclusion zone and proper protocol shall be implemented during Level B activities. The proper monitoring equipment, including, but not limited to, an OVA, PID, explosimeter, etc., shall be available on-site for all field activities when appropriate.

B1.6.2 Sampling Activity Equipment Needs

- **Level D Equipment**
  Leather boots, steel toe and shank;

  Hard hats;

  Safety glasses;

  Latex inner gloves; and

  Latex outer gloves.

- **Level C Equipment**
  Level D Equipment Plus:

  Hooded Tyvek suits;

  Outer boot covers; and

  Full face APR (NIOSH approved) with organic vapor canisters

- **Level B Equipment**
  Level C Equipment (with exception of APR) Plus:

  SCBA (positive - pressure, pressure demand, NIOSH approved)

Other equipment needs:

   Copy of site health and safety plan including a list of emergency contacts;

   First aid kit;

   Eye wash bottle;

   Paper towels;

   Duct tape;

   Drinking water; and

   Plastic garbage bags.

B1.6.3 Heat Stress

Adverse weather conditions are important considerations in planning and conducting site operations. Hot or cold weather can cause physical discomfort, loss of efficiency and personal injury. Of particular importance is heat stress resulting when protective clothing decreases natural body ventilation. Heat stress can occur even when temperatures are moderate. One or more of the following recommendations shall help reduce heat stress:

Provide plenty of liquids to replace body fluids (water and electrolytes) lost due to sweating.

Provide cooling devices to aid natural body ventilation. These devices, however, add weight and their use shall be balanced against worker efficiency.

Long cotton underwear acts as a wick to help absorb moisture and protect the skin from direct contact with heat absorbing protective clothing.

Install mobile showers and/or hose down facilities to reduce body temperature and cool protective clothing.

In extremely hot weather, conduct non-emergency response operations in the early morning or evening.

Ensure that adequate shelter is available to protect personnel against heat, cold, rain, snow or other adverse weather conditions which decrease physical efficiency and increase the probability of accidents.

In hot weather, rotate workers wearing protective clothing.

Good hygienic standards must be maintained by frequent change of clothing and showering. Clothing shall be permitted to dry during rest periods. Workers who notice skin problems shall immediately consult medical personnel.

Effects of Heat Stress

If the body's physiological processes fail to maintain a normal body temperature because off excessive heat, a number of physical reactions can occur. They can range from mild reactions such as fatigue, irritability, anxiety and decreased concentration, to severe reactions such as loss of dexterity or movement or death. Specific first aid treatment for mild cases of heat stress is provided in the American Red Cross first aid book. The location of this book shall be known at all times by the site manager and the book shall be readily available for reference in the field. Medical help must be obtained for the more serious cases of heat stress.

Heat related problems include:

Heat Rash: Caused by continuous exposure to heat and humid air and aggravated by chafing clothes. Decreases ability to tolerate heat as well as being a nuisance.

Heat Cramps: caused by profuse perspiration with inadequate fluid intake and chemical replacement, especially salts. Signs include muscle spasm and pain in the extremities and abdomen.

<u>Heat Exhaustion</u>: Caused by increased stress on various organs to meet increased demands to cool the body. Signs include shallow breathing; pale, cool, moist skin; profuse sweating; and dizziness and lassitude.

<u>Heat Stroke</u>: The most severe form of heat stress. Body must be cooled immediately to prevent severe injury and/or death. Signs include red, hot, dry skin; no perspiration nausea; dizziness and confusion; strong, rapid pulse; and possibly coma. Medical help must be obtained immediately.

B1.6.4 Air Monitoring Procedures

Air monitoring shall be used to identify and quantify airborne levels of hazardous substances. Periodic monitoring is required during all on-site activities. The flame ionization detector (OVA) shall be used to monitor ambient air concentrations in the worker breathing zone. A concentration of 1 ppm above background in the breathing zone shall necessitate the use of respirators. The explosivity meter shall be used to measure combustible gas levels at the well head before any activity begins at that well or in any confined spaces prior to entry. At 10 to 25 percent of the LEL implement the use of non-sparking tools and eliminate any flame or spark sources. at 25 percent of the LEL, evacuate the area and allow the area to ventilate.

B1.7 Site Control and Decontamination Procedures

The following site control measures shall be followed in order to minimize potential contamination of workers, protect the public from potential site hazards, and to control access to the sites. Site control involves the physical arrangement and control of the operation zones and the methods for removing contaminants from workers and equipment. The first aspect, site organization, is discussed in this section. The second aspect, decontamination, is considered in the next section.

B1.7.1 <u>Site Organization - Operation Zones</u>

The three operation zones established on the site are:

1. Exclusion Zone (contamination Zone)
2. Contamination Reduction Zone
3. Support Zone

The field team leader and/or site safety officer shall be responsible for establishing the size and distance between zones at the site or sub-sites.

Considerable judgement is required to assure safe working distances for each zone and are balanced against practical work considerations.

B1.7.2 <u>Exclusion Zone (Contamination Zone)</u>

The exclusion zone constitutes the place where active investigation or cleanup operations take place. Within the exclusion zone, prescribed levels of protection must be worn by all personnel. The hot line, or exclusion zone boundary, is initially established based upon the presence of actual wastes or apparent spilled material and is placed around all physical indicators of hazardous substances (i.e., drums, tanks, ponds, liquid run-off, defoliated areas). The hot line may be redefined based upon subsequent observations and measurements. This boundary shall be physically secure and posted or well defined by physical and geographic boundaries.

Under some circumstances, the exclusion zone may be subdivided into zones based upon environmental measurements or expected on-site work conditions.

B1.7.3 Contamination Reduction Zone

Between the exclusion zone and the support zone is the contamination reduction zone. This zone provides an area to prevent or reduce the transfer of hazardous materials which may have been picked up by personnel or equipment leaving the exclusion area. All decontamination operations are described Section B1.7.5 Decontamination Procedures.

B1.7.4 Support Zone

The support zone is the outermost area of the site and is considered a non-contaminated or clean area. The support zone contains the command post for field operation, first aid stations and other investigation and cleanup support. Normal work clothes are appropriate apparel within this zone; potentially contaminated personnel, clothing, equipment, etc., are not permitted.

B1.7.5 Decontamination Procedures

Personnel Decontamination Procedures

A decontamination area shall be set up and shall include provisions for collecting disposable protective equipment (such as garbage bags), washing boots, gloves, respirators (if used and field instruments and tools, as well as washing hands, face and other exposed body parts.

On-site personnel shall shower upon return to their homes at the end of the work day.

Decontamination equipment shall include:

    Plastic buckets and pails;

    Scrub brushes and long-handle brushes;

    Detergent;

    Containers of water,

    Paper towels;

    Plastic garbage bags; and

    Distilled water.

Decontamination of Equipment

All sampling equipment and tools shall be decontaminated according to standard decontamination procedures before removal from the site.

— Unit 2 —
Surface Water

Table of Contents

2.1	Regulatory Basis		58
	2.1.1	Colorado Surface Water Classification	58
		2.1.1.1 Recreational	58
		2.1.1.2 Agriculture	59
		2.1.1.3 Aquatic Life	59
		2.1.1.4 Domestic Water Supply	59
	2.1.2	Colorado Water Quality Standards	59
	2.1.3	Permits	60
	2.1.4	Storm Water Discharge	60
	2.1.5	Lead Contamination Control Act	61
2.2	Sampling Streams and Rivers		61
	2.2.1	Classification and Characteristics	61
	2.2.2	Sampling and Measurement	62
		2.2.2.1 Sampling Equipment	64
		2.2.2.2 Measurement Equipment	65
2.3	Sampling Lakes and Impoundments		67
	2.3.1	Classification and Characteristics	67
	2.3.2	Sampling	69
2.4	Sampling Drinking Water		72
	2.4.1	Sampling for Lead	72
	2.4.2	Asbestos in Water	73
2.5	Sampling Storm Water Runoff		74
2.6	Sources of Information		77
Appendix			77

— Unit 2 —

SURFACE WATER SAMPLING

2.1 REGULATORY BASIS

The basis for regulations concerning sampling and monitoring of surface water are found primarily in the Clean Water Act of 1972, the Safe Drinking Water Act (1974) and amendments to these acts known as the Lead Contamination Control Act of 1988 and the recent Storm Water Drainage Regulations under the Clean Water Act.

The Clean Water Act regulates surface waters through classification of surface waters, establishment of water quality parameters, and discharge limitations implemented through a permit system. Although the EPA has oversight authority to enforce provisions of the Clean Water Act, states are required to promulgate their own standards and classification systems, and to enforce the regulations.

2.1.1 Surface Water Classification

Most states classify surface waters according to the use for which they are presently suitable or intended to become suitable. For example, Colorado has classified all surface waters in the state except water in ditches and other manmade conveyance structures. Classifications are as follows:

2.1.1.1 Recreational

Class 1--Primary Contact

These surface waters are suitable or intended to become suitable for recreational activities (such as swimming or boating) in or on the water when the ingestion of small quantities of water is likely to occur.

Class 2--Secondary Contact

These surface waters are suitable or intended to become suitable for recreational uses on or about the water (such as fishing) which are not included in the primary contact classification.

2.1.1.2 Agriculture

These surface waters are suitable or intended to become suitable for irrigation or crops and livestock use.

2.1.1.3 Aquatic Life

Class 1—Cold Water Aquatic Life

These are waters that are currently capable of sustaining a wide variety of cold water (temperature rarely exceeds 20° C) biota or could sustain such biota but for correctable water quality conditions.

Class 1—Warm Water Aquatic Life

These are waters that are currently capable of sustaining a wide variety of warm water biota or could sustain such biota but for correctable water quality conditions.

Class 2—Cold and Warm Water Aquatic Life

Waters that are not capable of sustaining a wide variety of cold or warm water biota due to physical habitat, water flows or levels, or uncorrectable water quality conditions.

2.1.1.4 Domestic Water Supply

These are surface waters suitable or intended to become suitable for potable water supplies and can meet drinking water quality criteria following standard treatment.

2.1.2 Water Quality Standards

Although numerical water quality standards have been set, their application is complex and based on surface water classification, low stream flow considerations, the point of measurement as impacted by the mixing zone, and other factors. For example, Colorado has set standards for 6 radionuclides, 110 organic chemicals, 11 inorganic parameters, 19 metals, the biological parameter of fecal coliforms, and physical parameters including dissolved oxygen, pH, suspended solids, and temperature.

2.1.3 Permits

One important means of implementing water quality standards is through the National Pollutant Discharge Elimination System (NPDES). In essence, all point sources discharging wastes into surface water must have a NPDES permit. The permit itself sets allowable discharge limits in terms of amounts, concentrations, peak load variances, and any exemptions.

Waste characterization is a key part of the permit requirements. In most cases, data must be collected that represents current operations in terms of maximum daily values and average daily values. Grab samples must be used for pH, temperature, oil, grease, total chlorine, and fecal coliform. For other pollutants, 24-hour composite samples must be collected. In most cases each outfall or point discharge must be monitored.

The NPDES permit also covers the discharge of industrial wastes into the local sewer system. Most water treatment facilities cannot treat industrial wastes, and the wastes either travel through the plant untreated or disrupt the operation of the plant. Serious problems can result, such as the recent sewer explosions in Mexico where gasoline dumped into the sewer system exploded and killed over 200 people. To minimize problems with wastes discharged into sewers, industrial facilities are required to pretreat their wastes to remove specified substances before the waste enters the public sewer system.

2.1.4 Storm Water Discharge

According to the EPA, storm water is the last major uncontrolled source of water pollution. In response to this problem, the EPA requires NPDES permits for storm water discharged from large municipalities (serving a population more than 250,000 people), medium sized municipalities (serving a population from 100,000 to 250,000 people), for certain industrial facilities, and for construction sites over five acres. Permit requirements are similar to those mentioned above for the NPDES permit. Sampling and monitoring requirements include:

- quantitative data from at least 5 to 10 representative locations in approved sampling plans;

NOTES

- for selected pollutants, estimates of the annual pollutant loading and event mean concentration of system discharge;

- estimates of seasonal pollutant loads and the mean concentration of selected pollutants during a representative storm;

- field screening analysis for illicit connections and illegal dumping;

- characteristics and impacts of the waste water to the receiving system; and

- a proposed monitoring program for representative data collection.

2.1.5 Lead Contamination Control Act

In 1986 an amendment to the Safe Drinking Water Act requires the use of lead-free (less than 0.2 percent lead) materials in new plumbing and in plumbing repairs, and required water suppliers to notify the public about lead in drinking water. In 1988, the Lead Contamination Control Act became law. This requires the EPA to provide guidance to states to test for and remedy lead contamination in drinking water in schools and day care centers. In addition, proposed regulations would require that drinking water delivered by a public water system have lead levels equal to or less than 5 ppb as measured where the water leaves the treatment plant.

2.2 SAMPLING STREAMS AND RIVERS

2.2.1 Classification and Characteristics

Streams and rivers are conduits of surface water flow having defined beds and banks. Such water conduits are classified as follows:

Ephemeral

Water flows in ephemeral systems only in response to precipitation in the immediate watershed or in response to snow melt. The channel bottom is always above the local water table.

Intermittent

These are streams that commonly drain watersheds of at least one square mile and/or receive some of their flow from ground water during part of the year. However, intermittent streams do not flow continually.

Perennial

Perennial streams and rivers flow throughout the year in response to ground water discharge and/or surface water runoff.

Ephemeral and intermittent streams may not be obvious during periods of little or no precipitation. In many cases these streams are associated with topographic depressions in which surface water runoff is conveyed to receiving waters. Such areas may be distinctive because of the vegetation they support. For example, willows or cottonwood trees in a grassland indicate a supplemental source of water that is often classified as ephemeral or intermittent.

Perennial streams and rivers are continually engaged in a dynamic relationship with ground water. For any given stream segment, the perennial system is either receiving ground water discharge (and thus gaining water) or recharging the ground water and losing water.

The distinction between ephemeral, intermittent, and perennial will influence the selection of monitoring frequency, monitoring locations, and other design factors in the monitoring program. For example, the frequency of monitoring ephemeral streams, and to a lesser extent intermittent streams, will depend on rainfall runoff.

2.2.2 Sampling and Measurement

Important characteristics of streams and rivers that impact sampling include depth, velocity, turbulence, slope, changes in direction and cross section, and the nature of the bottom. These factors are often interrelated. For example, slope and roughness of the channel influence depth and velocity of flow, which together control turbulence. Turbulence affects rates of contaminant dispersion, reaeration, sedimentation, and rates of natural purification. All of these factors change from one stream segment to another and from season to season. Obviously, stream characteristics have a significant impact on the selection of monitoring sites.

Selection of monitoring locations is relative to the discharge under investigation. In most cases three areas should be sampled as follows:

Background Monitoring

Background monitoring stations should be located in an area known not to be influenced by the release; this location is usually upstream of the release.

Point of Release

If the release is a point source, periodic monitoring should be performed near the discharge point to determine the range of contaminant concentrations. The contaminant stream itself should also be monitored, generally at the outfall.

Area of Influence

The area of influence is defined as that portion of the receiving water affected by the discharge. This area is generally identified in a phased fashion, based on monitoring data. In most cases monitoring starts with a large area and as data become available, the boundaries are refined.

For long-term sampling, weirs or flumes are commonly installed at sampling points. A weir is a deliberate restriction inserted into a channel to force the water through a known cross section. Weirs can be triangular, rectangular, or trapezoidal in shape. A flume is a device placed in a channel to restrict the channel area and change the water flow, usually to obtain a constant head. Most flumes consist of a converging section to accelerate the approaching flow, a throat section whose width is used to designate flume size, and a diverging section to ensure that the downstream level is lower than the upstream converging level.

Monitoring frequency is a function of the type of contaminant release (intermittent or continuous), variability in water quality of the receiving water, stream flow, and other factors. Other factors in the sampling program relate to where in the water body the sample will be taken. Chemicals that float on water will require a discrete sampling of the surface layer of water. In contrast, insoluble chemicals heavier than water will require bottom sampling. Different sampling schemes are required for soluble vs. insoluble chemicals. Commonly used sampling and measurement equipment for streams and rivers are described below.

2.2.2.1 Sampling Equipment

Dipper

A dipper is an open-ended container attached to a pole. Commonly, the dipper and pole are made of Teflon and the pole is capable of telescoping for an extended reach. The dipper may be the actual sample container attached to a telescoping pole. This allows the sample to be retrieved, sealed, and prepared for transport without pouring the liquid from one container to another. Dippers are suitable for collecting samples of the surface layer of water or the water at a shallow depth below the surface. Due to the need of collecting a representative sample, the dipper should be used where the water is homogeneous; this most often occurs when water is moving rapidly, at a flume placed in the stream, or at an overflow device.

Depth Samplers *(i.e., Van Dorn Sampler)*

A variety of depth samplers are available where a closed sampling container is lowered into the stream and opened at a specific depth. Samplers are usually made of acrylic with a stainless steel closing mechanism. Specific depth samplers that may have application to streams and rivers are discussed in the following section on lakes and impoundments.

Bailers and Pumps

In some situations, a well bailer can be used to collect water from streams and rivers; details concerning bailer type and use are presented in Unit 3. Also, portable pumps, particularly a peristaltic pump, can be used for sample collection. Flexible tubing (such as Teflon, polypropylene, etc.) is used. Most sampling procedures call for pumping several liters of sample through the system before collecting a sample. However, the use of peristaltic pumps for sampling VOCs is not recommended because the squeezing action of the pump can strip VOCs from the water. Additionally, silicone and neoprene tubing is not generally recommended for sampling of organics.

2.2.2.2 Measurement Equipment

Turbidity

Turbidity, or the amount of suspended and often dissolved substances in water, is measured by light scattering techniques. Water is placed in an optically selected sample vial and placed in a light-free chamber. A Tungsten lamp is then activated and a photovoltaic photodetector records the light scattered at right angles from the beam of light passing through the sample. Units of measurement are Nephelometric Turbidity Units or NTUs.

Dissolved Oxygen

A variety of dissolved oxygen meters are available that give direct readings of oxygen concentrations in surface waters. Most units use a probe that is lowered by a cable to the desired sampling depth. Many units are calibrated for changes in temperature and altitude.

Current Velocity

Most systems have a mechanical device with a rotating element that, when submerged, rotates at a speed proportional to the velocity of the flow. Each revolution creates an electronic pulse that is processed and displayed digitally in feet or meters per second. Another type of current meter uses electromagnetic sensors where the passage of fluids between two electrodes in a bulb-shaped probe causes a disturbance of the electromagnetic field surrounding the electrodes. This disturbance generates a small voltage that is proportional to the fluid velocity.

Water velocity varies by depth and location within the channel. Typically, water is flowing slower on the surface and at the bottom (due to friction) than in the middle portion of the stream. Velocity also changes from one side of the stream to the other. Therefore, measurements must be made at numerous depths and locations and integrated to accurately determine the overall velocity.

In addition to velocity, the cross-sectional area of the stream must be determined. This is normally done by using measuring tapes and weighted chains or lines. When the cross sectional area and velocity are multiplied, the flow rate, in units of cubic feet/second or gallons/minute for slower flows, is obtained.

In addition to these chemical and physical parameters, biological measurements are also used to evaluate the quality of streams and rivers. Of particular interest are aquatic macro-invertebrates. These organisms spend the greatest portion of their life cycle in water, generally as nymphs, naiads, or larvae. The adults leave the aquatic environment as winged insects, mate, and deposit eggs in or near the water. The quality of water has a significant impact on the species present and their abundance. Macro-invertebrates can be collected on plate samplers that consist of hardboard plates (one side smooth and one side rough) fastened together and inserted into the water for approximately two weeks. Fine-meshed nets can also be used for collection. Macro-invertebrates are identified using a microscope.

Techniques to keep in mind when sampling are as follows:

When sampling surface water, the sampler should stand down stream and face up stream to minimize disturbance. The mouth of the sample container should also face upstream.

Care must be taken when collecting floaters because these substances can adhere to the sample container and only a portion of the floaters may be transferred into the sampling bottle for analysis.

Keep in mind that streams and rivers are not homogeneous from bank to bank and numerous samples and/or measurements must be taken to accurately characterize the stream. All sampling points should be marked in the field and sketched on a map.

2.3 SAMPLING LAKES AND IMPOUNDMENTS

2.3.1 Classification and Characteristics

Lakes and ponds are generally considered natural features while lagoons, reservoirs, and other types of impoundments are humanmade. In both cases, the source of water is either surface water (including discharged waste water), ground water, or both. As with streams and rivers, the physical characteristics of lakes and impoundments influence the transport of contaminants and therefore the design of the monitoring program. Important physical characteristics include dimensions (length, width, shoreline, and depth), temperature distribution, and flow pathways.

Temperature distribution is especially important in lakes and impoundments where vertical temperature stratification is found. Three temperature zones (Figure 1) are identified as follows:

Epilimnion

This is the surface layer of water and is characterized by good light penetration, high levels of dissolved oxygen, low levels of carbon dioxide, good mixing due to wave action, and elevated biological activity.

Mesolimnion (Thermocline)

This is generally a narrow zone between the epilimnion and hypolimnion and is characterized by rapidly changing temperatures, moderate levels of oxygen and carbon dioxide, and little mixing and biological activity.

Hypolimnion

This zone is found at the bottom of the lake or impoundment and is characterized by poor light penetration, low levels of dissolved oxygen, high levels of carbon dioxide, poor mixing, and little biological activity.

During most of the year, these layers are distinctive and little mixing occurs from layer to layer. However, mixing occurs during "overturn" in temperate climates. Water reaches its greatest density at 4 degrees Centigrade; both above and below this temperature water is less dense. During the fall, the epilimnion

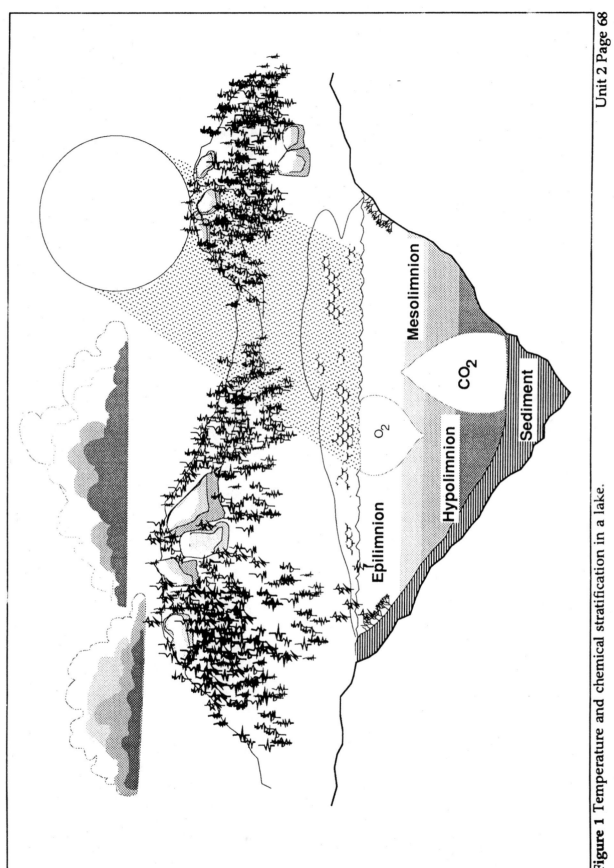

Figure 1 Temperature and chemical stratification in a lake.

cools and becomes denser than the underlying strata. At some point, the underlying strata can no longer support the denser, heavier water and the lake overturns, resulting in mixing. A similar phenomenon occurs in the spring as the surface water warms to 4 degrees Centigrade and once again become denser than the underlying waters.

Stratification is important in the transport of contaminants within a lake or impoundment, and the location of monitoring points in large bodies of water is significantly influenced by temperature zones. In contrast, for smaller bodies of water that are not significantly stratified, the location of monitoring points is strongly influenced by horizontal flow paths, shoreline configuration, and other factors.

2.3.2 Sampling

Due to stratifications and the potential for certain contaminants to concentrate in specific depth zones, most sampling plans specify that lakes and impoundments be sampled by depth. Therefore, equipment capable of collecting a sample at a specific depth is commonly used (Figure 2). A description of these discrete sampling devices follows.

Subsurface Grab Sampler

A bottle is attached to a valve assembly and both units attached to a pole. The sampler is lowered to the desired depth and a cable on the pole pulled. The cable opens the valve and the bottle fills as the air escapes. A modular pole design allows for the attachment of six foot long sections up to a total length of 30 feet.

The pole for the sampler is made of Teflon or stainless steel, the valve fitting is polypropylene or Teflon, and the sampling container is glass.

Kemmerer Sampler

This instrument is a spring-loaded cylinder that is lowered into a liquid in the open position to allow the liquid to flow through it while the sampler is descending. At the desired depth, a messenger is dropped down the line, releasing a spring-loaded closing device that seals the sample in the container. Samplers are made of brass,

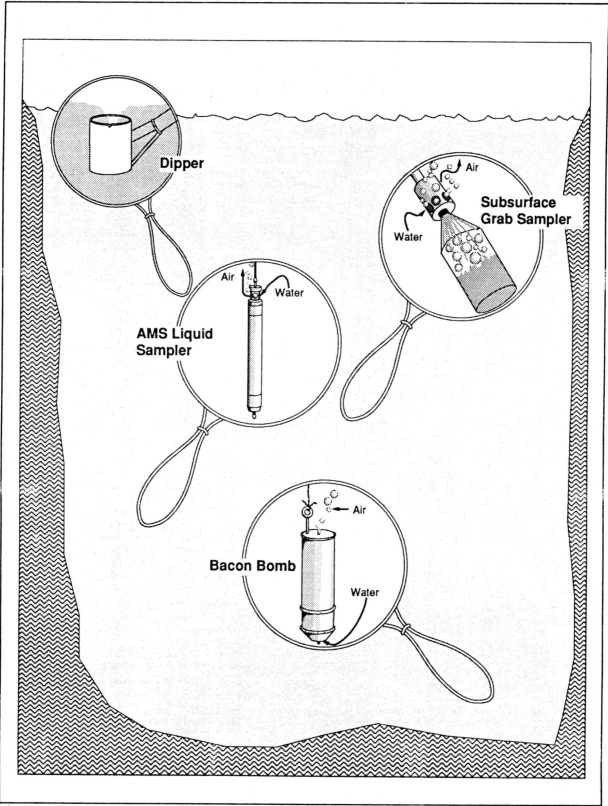

Figure 2 Sampling devices for surface water.

Teflon, or stainless steel. One disadvantage of the Kemmerer Sampler is that it descends in the open position; therefore, contaminants in the upper portion of the lake can adhere to the sampler and can be incorporated into a sample taken at a deeper depth.

Bacon Bomb Sampler

This sampling device, which looks like an average, all-purpose bomb, is lowered in a closed position to the desired sampling depth on a cable. An inner rod is then pulled that opens the sampler. A discrete-depth sample can be collected with a minimal amount of agitation since the opening rod runs down the center of the sampler and both ends open at once, thus allowing water to enter through the bottom and air to exit through the top. Samplers are made of brass, Teflon, or stainless steel.

AMS Liquid Sampler

This stainless steel and Teflon sampler is designed to obtain a discrete sample at a specified depth or a composite sample from several intervals. It consists of a sampling cylinder with a spring-loaded ball valve (or other type of valve) on the top. When the desired sampling depth is reached, a stainless steel cable is pulled, thus opening the valve and collecting a water sample in a manner similar to the subsurface grab sampler. For shallower depths, a stainless steel rod is available rather than a cable. The sample is released by a spring-loaded plunger on the bottom.

Other Instruments

A well bailer can be used to collect samples from lakes, but it is not often used because of water exchange during the ascent. Instruments used to collect water quality data for lakes and impoundments are similar to those described for streams and rivers, with exception of the current meter that is not applicable to standing bodies of water.

Decontamination of discrete depth samplers can be difficult because of internal valves and cables. Equipment selection should consider the case of decontamination.

Monitoring locations and sampling frequency follow the same general rules as mentioned for streams and rivers. In addition to depth considerations, samples are often collected from shore to shore in cross section fashion to account for horizontal variability in lakes and impoundments.

2.4 SAMPLING DRINKING WATER

Strict quality control measures for public drinking water have restricted the sampling of drinking water to situations where there is some concern with the privately installed delivery system from the water main to the tap. Currently, the corrosion of lead from the delivery system is of concern. In addition, where asbestos bearing transite pipes have been used for the delivery system, asbestos in drinking water is of concern. Following is the general sampling protocol for both lead and asbestos in public drinking water supplies.

2.4.1 Sampling for Lead

Most sources of drinking water have no lead or very low concentrations of this metal. However, lead can readily contaminate water through the water supply system. Corrosion of lead pipes, solder, fixtures, and other portions of the plumbing system is the most common source of lead in drinking water. The corrosion is caused by "soft" water (water containing a low concentration of ions) and acidic water. Because lead in plumbing systems was outlawed in 1986, water systems built after that date have a low probability of lead contamination.

Another source of lead is from lead-lined water coolers. A 1989 law requires that these coolers must be repaired to remove the lead, replaced, or recalled in schools and child care centers.

The recommended EPA sampling protocol is as follows:

> Collect all samples before school opens and before any water is used. Ideally, the water should sit in the pipes unused for at least 8 hours but not more than 18 hours before sampling.
>
> For sampling the service connection (the pipe connecting the main line to the facility), select the tap closest to the service connector. Let the water run and feel the temperature of the water. As soon as the water changes

from warm to cold, collect the sample. Because water warms slightly after standing in the interior plumbing, this cooler water represents the water that has been standing outside the building and in the service connector. Fill a pre-cleaned plastic container with 250 ml of water.

For sampling water in the water main, let the water run until it changes from warm to cold. Then let the water run for an additional 3 minutes to purge the service connection. Following purging, collect a sample using a 250 ml pre-cleaned plastic container.

For water fountains, the objective is to collect a sample that is representative of the water consumed at the beginning of the day or after infrequent use. It consists of water that has been in contact with the fountain valve and fittings and that section of plumbing closest to the outlet of the unit. This sample is taken before school opens and before any water is used. Collect the water immediately after opening the faucet and without allowing any water to run down the drain.

2.4.2 Asbestos in Water

Prior to sampling for asbestos from a faucet, all hoses and other non-plumbing attachments should be removed. The water should then run to purge the pipes of stale water.

A 1-liter sample container composed of glass or low density polyethylene should be used to collect the water sample. Other materials should be avoided due to possible particulate contamination. The bottle should first be rinsed twice by filling it approximately one-third full with fiber-free water and shaking it vigorously for 30 seconds. Glass bottles can have significant levels of asbestos on the interior surface and should be thoroughly rinsed prior to use.

Following rinsing, approximately 800 ml of water should be collected. Air space must be left in the bottle to allow for redispersal of settled material before analysis. Preservatives are not required in the field. However, in the laboratory, the sample must be preserved with ozone-UV treatment or mercury compounds within 48 hours after collection to prevent bacterial growth.

2.5 SAMPLING STORM WATER RUNOFF

The storm water runoff regulations specify the type of quantitative data required from each storm. Samples must be collected from a "storm event" defined as greater than 0.1 inches of precipitation that occurs at least 72 hours after the last storm event. In addition, the duration and total rainfall for the storm event should be from 0.5 to 1.5 times the average storm event for the area being monitored. The data requirements for a storm event include maximum flow rate, total volume of discharge, date, duration and amount of precipitation, and elapsed time after the last storm event.

Two types of samples must be collected. A "first flush grab sample" is defined as a sample taken during the first 30 minutes of the storm event. Any pollutants in this sample can be used as a screen for non-storm water discharges since such pollutants are flushed out of the system during the initial portion of the discharge. A "flow-weighted composite sample" is a composite sample taken for the entire storm event. This sample is composited by combining many small samples of at least 100 ml each. The time interval between samples, or the volume of each sample, is proportional to the total flow since the last sample was taken. Each sample type is analyzed separately for pollutants.

For industrial dischargers, first flush grab samples and flow-weighted composite samples must be analyzed for biological oxygen demand, chemical oxygen demand, total suspended solids, total phosphorus, total nitrogen, and specific pollutants. In addition, the first flush grab sample must be analyzed for oil, grease, and pH. For municipalities, analyses include those listed above and total dissolved solids, fecal strepto coccus and coliform, dissolved phosphorus, total ammonia, specified metals, volatile organic compounds, acids, bases, and pesticides.

To summarize, requirements for storm water runoff sampling must monitor rainfall, measure flow, and take first flush grab and flow-weighted composite samples. Monitoring systems should also record rainfall events, flow, and sampling times. Integrated systems that can be programmed to automatically record measurement and collect samples once a storm event occurs are commercially available. For example, an integrated system may work as follows:

NOTES

A flow meter monitors rainfall signals from a rain gauge and from the flow rate in a discharge channel. The flow meter activates the sampler when a predetermined amount of rain falls in a predetermined time period and the flow rate in the discharge channel rises to a predetermined level. The flow meter then sends signals to the sampler to collect samples (Figure 3). The sampler collects separate first flush samples and flow-weighted composite samples. The flow rate is plotted on an internal flow meter chart along with sample data and hourly rainfall amounts.

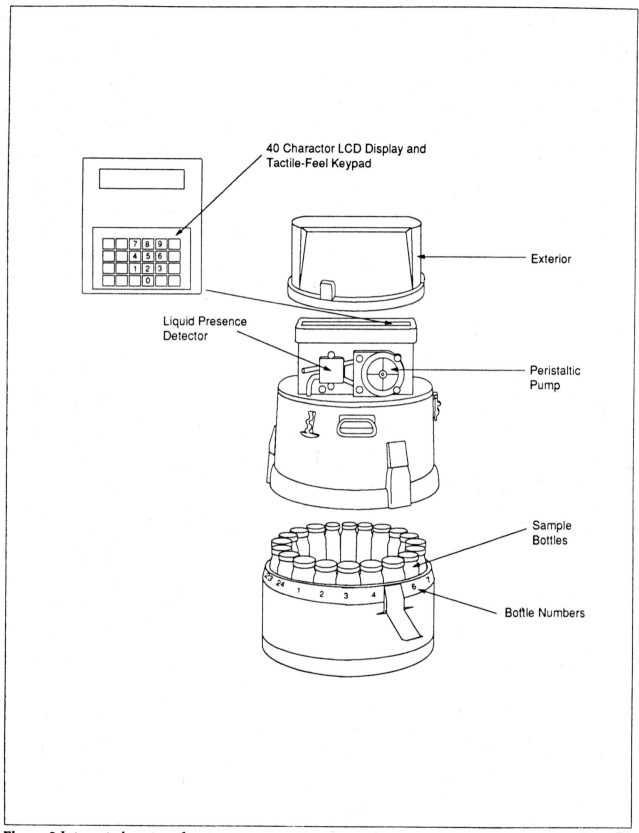

Figure 3 Integrated system for storm water and waste water sampling.

2.6 SOURCES OF INFORMATION

Environmental Protection Agency. 1989. Lead in School Drinking Water. EPA 570/9-89-00.

Environmental Protection Agency. 1987. A Compendium of Superfund Field Operation Methods. EPA 540/P-87-001.

Environmental Protection Agency. 1989. Interim Final RCRA Facility Investigation (FRI) Guidance, Volume III of V, Air and Surface Water Releases. EPA 530/SW-89-031.

Gordon, N.D., T.A. McMahon, and B.L. Finlayson. 1992. Stream Hydrology. John Wiley & Sons, New York.

Wanielista, M.P. and Y.A. Yousef. 1993. Stormwater Management. John Wiley & Sons, New York.

Appendix

Indicator Parameters

Following is a discussion of commonly used parameters that are often used to indicate the quality of water.

Biochemical Oxygen Demand (BOD)

BOD is a measure of the dissolved oxygen consumed by microbial life while assimilating and oxidizing organic matter in water.

Chemical Oxygen Demand (COD)

COD measures the amount of oxygen consumed under specific conditions in the oxidation of organic and oxidizable inorganic matter in water.

Total Organic Carbon (TOC)

TOC is used to measure the carbon present in water as part of organic compounds. However, this measurement is not specific to a given contaminant or even to a specific class of organics.

Dissolved Oxygen (DO)

This is the oxygen dissolved in water and available for use by aquatic organisms. DO levels less than 3 mg/liter are considered stressful to most aquatic vertebrates. Dissolved oxygen can be depleted by a variety of means, such as high BOD values.

pH

This parameter is the negative logarithm of the hydrogen-ion activity. It determines the acidity (pH less than 7.0) or alkalinity (pH greater than 7.0) of the water. The value of pH is in its predictive ability, especially in terms of inorganic constituents. When pH is low (less than approximately 5.0), metals are likely to be found dissolved in the water. When pH is high (greater than approximately 8.0) salts are likely to be found in the water. Most organic substances are near neutral (pH of 7.0) and pH measurements often do not supply useful information about organic products.

Electronic instruments are used to measure pH. This parameter should be measured immediately after the sample is collected. pH changes with temperature and with sample preservation solution, so immediate field measurement is necessary. Most instruments utilize both a glass electrode and a reference electrode. The tips of the electrodes are inserted into the water and the pH recorded once the readings have stabilized.

Temperature

Water temperature is a fundamental parameter that should be recorded whenever water samples are collected. This parameter can be significant because:

- most aquatic organisms are sensitive to elevated temperatures;

- elevated temperatures can be an indication of contamination;

- most chemical reactions are temperature dependent; and

- temperature defines zones in thermally stratified water bodies.

Alkalinity

Alkalinity is the capacity of water to resist a depression in pH or to accept hydrogen ions without decreasing the pH. It is usually expressed in calcium carbonate equivalents and is the sum of alkalinities provided by the carbonate, bicarbonate, and hydroxide ions present in the solution. Alkalinities in the natural environment usually range from 45 to 2000 mg/liter.

Hardness

Hardness is the measure of the total concentration of calcium and magnesium present in solution, expressed as calcium carbonate equivalents. Hardness is a desirable quality in water because both calcium and magnesium are nutrients.

Total Solids (TS)

This is a measure of the solids that remain after the water has evaporated; this residue consists of suspended, colloidal, and dissolved solids.

Suspended Solids (SS)

Suspended solids are those organic and inorganic materials that will not pass a glass-fiber filter. As a point of reference, domestic waste water generally contains 2000 mg/l of suspended solids.

Total Dissolved Solids (TDS)

Total dissolved solids (TDS) is a measure of the concentration of dissolved inorganic ions and the relatively high concentration of organic ions in water. Units of measurements are grams/liter or milligrams/liter (mg/l). For inorganic ions, TDS is another means of expressing conductivity, and the two parameters are functionally related.

Specific Conductance

Conductivity is the ability of a solution to conduct an electric current. The current is carried by inorganic ions. Ions with a negative charge are called anions and those with a positive change are termed cations. Examples of common anions in water include nitrate, phosphate, chloride, and sulfate. Common cations found in water include sodium, calcium, iron, magnesium, and aluminum. However, conductivity is not ion-specific, but measures the sum total of ions (as determined by electrical conductivity) in solution. Organic material such as alcohols, oils, phenols, and sugars do not carry an electric current well and do not have sufficient conductivity for a detectable measurement.

Conductivity is measured by applying a voltage between two electrodes immersed in the test solution. The voltage drop caused by the resistance of the solution is determined and converted to its reciprocal conductivity. The basic unit of measurement is siemens/cm or mhos/cm (there is no conversion factor). Because of the conductivity range normally found in water, millisiemens/cm (1/1000 of a siemen) is commonly used. Most conductivity probes are temperature compensated to the reference temperature of 25 degrees Celsius.

Conductivity is readily measured in the field by using an electronic probe with a readout unit. This parameter can be related to the chemical purity of water, the amount of dissolved salts in water, salt concentration in brine, and other characteristics.

Turbidity

Turbidity instruments measure the suspended solids in water. Suspended solids can be silt and clay particles or precipitated chemicals such as iron oxide. For most instruments, a sample of water is placed in a Pyrex cell and inserted into a chamber. The chamber is sealed from outside light and a light source within the chamber is activated. Turbidity of the sample scatters some of the light; this scattered light is sensed by a photocell and transmitted to a readout unit. Turbidity is measured in terms of NTU (nephelometric turbidity units).

— Unit 3 —

Ground Water Sampling

Table of Contents

3.1		Federal Regulations Affecting Ground Water Monitoring	82
	3.1.1	Resource Conservation and recovery ACT (RCRA)	82
		3.1.1.1 Subtitle C--Hazardous Waste Facilities	82
		3.1.1.2 Subtitle D--Non-Hazardous Solid Waste Disposal Facilities	83
		3.1.1.3 Subtitle I--Underground Storage Tanks	84
	3.1.2	Comprehensive Environmental Response, Compensation, and Liability Act (CERCLA)	84
	3.1.3	Clean Water Act	85
	3.1.4	Safe Drinking Water Act (SDWA)	85
	3.1.5	Surface Mining Control and Reclamation Act (SMCRA)	85
3.2		Ground Water Quality Standards	85
3.3		Characteristics of Ground Water	87
	3.3.1	The Saturated Zone	87
	3.3.2	Ground Water Flow	90
	3.3.3	Aquifer and Other Terms	93
3.4		Ground Water Monitoring Wells	94
	3.4.1	Number and Placement	94
	3.4.2	Product Density and Solubility	96
	3.4.3	Selection of the Monitoring Zone	96
	3.4.4	Drilling	99
		3.4.4.1 Drilling Without Circulating Fluids	100
		3.4.4.2 Drilling With Circulating Fluids	103
		3.4.4.3 Selection of the Appropriate Drilling Method	106
	3.4.5	Permitting	107
	3.4.6	Well Design and Instillation	107

(continued on next page)

Table of Contents

		3.4.6.1	Well Diameter	107
		3.4.6.2	Well Casing	109
		3.4.6.3	Screening	111
		3.4.6.4	Filter Packs	113
		3.4.6.5	Annular Seals	116
		3.4.6.6	Surface Protection	117
	3.4.7	Temporary Ground Water Wells		119
	3.4.8	Decontamination		119
3.5	Well Development			120
	3.5.1	Flushing and Surging		120
		3.5.1.1 Overpumping		121
		3.5.1.2 Backwashing		121
		3.5.1.3 Mechanical Sluging		122
		3.5.1.4 Air Surging		122
		3.5.1.5 Water Jetting		123
	3.5.2	Removal of Debris and Contaminants		123
	3.5.3	Slug Testing		123
	3.5.4	Surveying		126
3.6	Ground Water Sampling			126
	3.6.1	Well Approach		126
	3.6.2	Well Purging		127
		3.6.2.1 Casing Volume Criterion		127
		3.6.2.2 Parameter Stabilization Criterion		128
		3.6.2.3 Hydraulic & Parameter Stabilization Criterion		128
		3.6.2.4 Purging Rate		128
		3.6.2.5 Purging Volume		129
		3.6.2.6 Purging Location		129
		3.6.2.7 Purge Reduction		130
	3.6.3	Determining Casing Volume		133

(continued from previous page)

Table of Contents

	3.6.4	Sampling Equipment	136
		3.6.4.1 Bailers	136
		3.6.4.2 Ground Water Pumps	138
	3.6.5	Filtration	145
		3.6.5.1 Vacuum Filters	146
		3.6.5.2 Pressure Filters	146
		3.6.5.3 Disposable In-Line Filter Cartridges	148
		3.6.5.4 In-Line 142mm Backflushing Filter	148
3.7	Decontamination		148
3.8	Quality Assurance and Quality Control		151
3.9	Sources of Information		152
Appendix A	Colorado Well Permit Form		154
Appendix B	Colorado Well Construction and Test Report		159
Appendix C	Forms Commonly Used in Well Development and Sampling		162
Appendix D	Colorado Well Closure Form		167

— Unit 3 —

GROUND WATER SAMPLING

Ground water sampling is an integral part of most hazardous waste site investigations. The investigation includes a number of carefully planned and executed steps including placement, drilling, development, maintenance, and sampling of the ground water wells. The specific hydrogeological conditions of a site strongly influences each of these factors. Consideration must be given to the types and distribution of geologic materials, the occurance and movement of ground water through these materials, the location of the site in the regional ground water flow system, and the relative permeabilities of the materials. Potential interactions between the contaminants or analytes of interest and the geochemical and biological constituents of the formation(s) must also be considered.

3.1 FEDERAL REGULATIONS AFFECTING GROUND WATER MONITORING

Following is a brief summary of the major federal regulations that mandate ground water monitoring; many of these programs are state implemented. Additionally, most states have ground water monitoring requirements that supplement the federal programs.

3.1.1 Resource Conservation and Recovery Act (RCRA)

3.1.1.1 Subtitle C--Hazardous Waste Facilities

Subtitle C addresses the management of hazardous waste from its generation to it ultimate disposal. Various sections of these far-reaching regulations address ground water monitoring. Part 264 addresses owners and operators of permitted hazardous waste treatment, storage, and disposal facilities (TSDF). Three phases of ground water monitoring may be required.

- Phase I. Virtually all TSDF must have a ground water monitoring scheme where a sufficient number of wells are installed at appropriate locations and depths. Samples from the upper most aquifer must

(1) represent the quality of ground water that has not been affected by the facility and (2) the quality of ground water that may be affected by the facility.

- Phase II. If statistical analysis of the data generated by the above wells indicates a release from the facility, the next level of ground water monitoring must be implemented. This phase is designed to fully characterize the nature and extent of ground water contamination.

- Phase III. Here the owner or operator selects and implements approved corrective actions to restore ground water quality. Part of this is the Compliance Monitoring Program where the primary purpose is to monitor ground water to determine the effectiveness of the corrective action program.

Further ground water monitoring is required during closure and post-closure periods. If all the wastes are not removed from the facility, this post-closure monitoring period could last 30 years.

3.1.1.2 Subtitle D--Non-Hazardous Solid Waste Disposal Facilities

Regulations issued in 1991 require that municipal solid waste landfills (MSWLF) implement, by 1996 for existing facilities, a ground water program similar to the Subtitle C program. Three phases of monitoring are required as follows:

- Phase I. The ground water monitoring system must consist of a sufficient number of appropriately located wells that yield ground water samples from (1) the uppermost aquifer that represents the quality of background ground water and (2) indicate the quality of ground water passing the relevant point of compliance at the MSWLF. Ground water must be sampled semiannually and analyzed for specified parameters.

- Phase II. If any of the monitoring parameters are statistically higher than background values, the owner or operator must expand the monitoring program to characterize the nature and extent of contamination.

- Phase III. The owner or operator must select and implement a corrective measures plan that includes a corrective action ground water monitoring program. Following compliance, the monitoring program must continue for three consecutive years.

In addition, closure and post-closure requirements specify a ground water monitoring program that may last 30 years following closure.

3.1.1.3 Subtitle I–Underground Storage Tanks

As presented in a later section, Subtitle I does not specifically require ground water monitoring; it is listed as one option for the required tank monitoring program. Many states have specific language that requires UST owners or operators to install ground water monitoring wells adjacent to each new UST.

Confirmed releases from USTs requires a corrective action plan that addresses remediation of soil and ground water, as required. If ground water has been contaminated, a monitoring program must be implemented to determine the effectiveness of corrective actions.

3.1.2 Comprehensive Environmental Response, Compensation, and Liability Act (CERCLA)

Ground water monitoring is integrated throughout the CERCLA program. In most situations, monitoring starts during the initial investigation to support site characterization and to provide the data required for the Hazard Ranking System (HRS). Once a site has been placed on the National Priority List, ground water monitoring continues and is often greatly expanded in the Remedial Investigation (RI) where pathways for and receptors of hazardous substances are quantified. Monitoring may continue during the Feasibility Study (FS) to assist in the evaluation of remedial technology. Ground water monitoring is also a critical factor in evaluating the effectiveness of corrective actions. Following site closure, monitoring continues for an indeterminate time.

CERCLA has not established any specifications for ground water monitoring. Rather, each site is evaluated individually and a specific monitoring program established.

3.1.3 Clean Water Act

Although this act focuses on the protection of surface water, ground water monitoring may be required in some cases. For example, if ground water and surface water are hydraulically connected as is often the case, the protection of surface water quality will require protecting ground water quality.

In some cases, specific ground water monitoring programs are required. For example, if federal funds are used to construct a municipal sewage treatment plant that will apply waste water to land, ground water monitoring is required.

3.1.4 Safe Drinking Water Act (SDWA)

This act protects ground water through the establishment of drinking water standards, sole source aquifer designation, the Well Head Protection Program, and the Underground Injection Control Program. Although all of these programs require some degree of ground water monitoring, the Underground Injection Control Program requires the most intensive monitoring. Under this program, ground water monitoring is required to evaluate whether an underground source of drinking water may be endangered by injection of fluids that are divided into five classes of wells. The type of monitoring wells, their number and location, sampling frequency, and the parameters to be measured are specified on a site-by-site basis.

3.1.5 Surface Mining Control and Reclamation Act (SMCRA)

As part of the reclamation program specified by this act, baseline ground water information must be collected and compared to ground water quality in reclaimed and adjacent areas.

3.2 GROUND WATER QUALITY STANDARDS

The results of ground water monitoring programs must be compared with some useful reference point. Some of these reference points, or quality standards, are discussed below.

In 1984, the EPA established the Federal Ground Water Protection Strategy that provided for differential protection of ground water depending on its resource value. Three classes of ground water were established as follows:

- Class I--ground water that is highly vulnerable to contamination and is either an irreplaceable source of drinking water or is ecologically vital.

- Class II--ground water that is a current and potential source of drinking water and ground water having other beneficial uses, such as irrigation.

- Class III--ground water that is heavily saline or otherwise contaminated beyond the level allowing cleanup through methods commonly used by public water supply treatments.

The SDWA has set standards that are applied to ground water when it is used as a source of drinking water. Two levels have been set as follows:

- Primary Drinking Water Standards--this applies to public water systems and specifies maximum contaminant levels (MCLs) for water or a reduction level based on water treatment.

- Secondary Drinking Water Standards--this advisory standard applies to public water systems and specifies limits on physical characteristics that may not affect health, but may make water less pleasing to drink or use due to color, odor, staining, foaming, and cloudiness.

The MCLs are enforceable standards and apply to ground water that is used for both public and private drinking water. In setting the standards, best available technology and cost are considered. In addition, maximum contaminant level goals (MCLG) have been established and represent the maximum level of a contaminant in drinking water at which no known or anticipated adverse effect on human health would occur, allowing for an adequate safety margin. These goals are not enforceable.

The CWA has set water quality criteria known as the 304(a)(1) criteria. These criteria are not enforceable and are established to provide guidance on the environmental effects of pollutants. Although derived for surface water, they are often applied to ground water.

Selection of appropriate ground water quality standards for a particular situation is neither simple nor straightforward. A variety of factors and standards must be evaluated by a professional staff in consultation with the appropriate regulatory agencies.

3.3 CHARACTERISTICS OF GROUND WATER

Ground water is an integral component of the global hydrologic cycle (Figure 1). This cycle provides for the continuous circulation of water from land and sea to the atmosphere and back again. Water evaporates from oceans, lakes, rivers, soil, and plants into the atmosphere. This water eventually precipitates as rain or snow onto the land where it evaporates, runs off into streams and rivers, or infiltrates into the soil from which some is transpired back into the atmosphere by plants. The remaining water in the soil moves downward into geologic formations and becomes ground water that eventually seeps into streams or lakes from which it evaporates, or flows to the oceans.

The hydrologic cycle demonstrates that water is always in motion. Surface and atmospheric water move at a rapid rate, often measured in feet per second. In contrast, the movement of ground water is slow and is usually measured in feet per year. In general, near-surface ground water moves from approximately 75 to 150 feet per year.

3.3.1 The Saturated Zone

Water and any associated contamination pass through several different hydrologic zones as they migrate to ground water (Figures 1 and 2). The soil and subsoil zones are characterized by predominantly unsaturated conditions; that is, the void space in the soil is partially filled with water and partially filled with air. Overall direction of water movement is downward, although substantial upward water movement occurs near the soil surface during dry conditions.

A water table is found at some distance below the surface; this distance can range from less than a foot to hundreds of feet depending on the climate and subsurface conditions. The water table is the surface of the ground water. Below the water table saturated conditions exist and all pores are filled with water. However, there are some exceptions. Certain rock configurations may trap gasses or isolated voids may exist that do not fill with

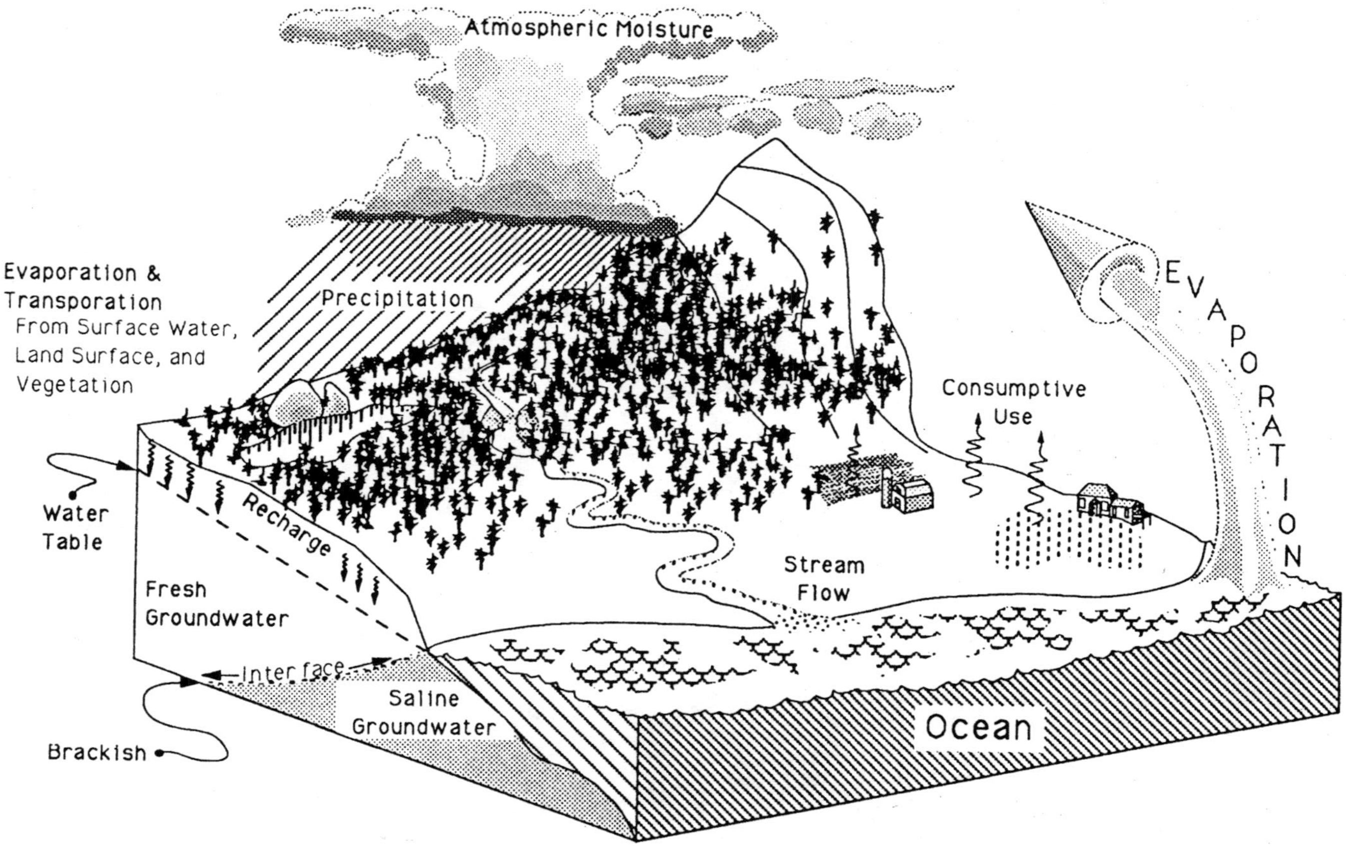

Figure 1 Water in the environment.

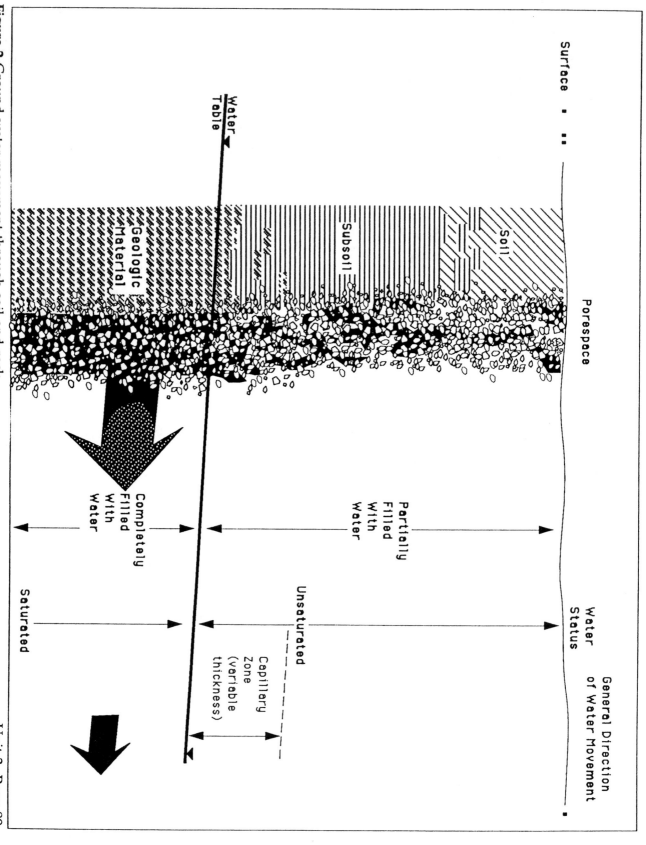

Figure 2 Ground water movement through soil and rock.

water. General direction of water movement in this zone is usually lateral. Depth to the water table will fluctuate by seasons and year as recharge and discharge change. Generally, the shallower the water table, the greater the seasonal and annual fluctuations.

Immediately above the water table is the capillary zone (fringe). In this zone liquids can move upward, against the force of gravity, in response to the forces of adhesion and cohesion. The liquid forms a film around particles, filling the interstitial (void) spaces, and moves from particle to particle in a process similar to water moving upward in blotter paper. This zone is saturated near the water table and gradually decreases in water content with increasing elevation above the water table. Depth of this zone can range from several inches to several feet, depending primarily on soil texture. Liquid movement in the capillary zone is upward, downward, and horizontal.

3.3.2 Ground Water Flow

The manner in which ground water flows through an aquifer is dependent upon the hydrogeologic conditions and influential artifacts (e.g. water impoundment). There are three basic types of geologic materials normally encountered in ground-water monitoring programs: 1) porous media, 2) fractured media, and 3) fractured porous media. In porous media, the water and contaminants move through the pore spaces between individual grains of the media. These media include sand and gravel, silt, loess, clay, till, and sandstone. In fractured media, the water and contaminants move through cracks or solution crevices in otherwise relatively impermeable rock. These media include dolomites, some shales, granites, and crystalline rocks. In fractured porous media, the water and contaminants move through both the intergranular pore spaces and the cracks or crevices in the rock or soil. The occurrence and movement of water through the pores and cracks or solution crevices depend on the relative porosity and degree of channeling from cracks or crevices. These media include fractured tills, fractured sandstone, and some fractured shales. Below is an illustration of the occurance and movement of water and contaminants in these three types of geologic materials.

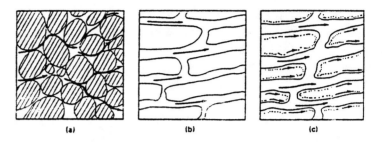

Occurrence and movement of ground water through
a) porous media, b) fractured or creviced media, c) fractured porous media.

The distribution of these three basic types of geologic materials is seldom homogeneous or uniform. In most settings, two or more types of materials will be present. Even for one type of material at a given site, large differences in hydrologic characteristics may be encountered. The heterogeneity of the materials can play a significant role in the rate of contaminant transport, as well as in the selection of the optimum strategy for monitoring a site.

In the above diagrams, the flow of ground water is, for the most part, along distinct and separate paths that are generally parallel to one another. This is called laminar flow. In laminar flow, there is very little mixing of water and associated contaminants. In contrast, the flow of surface water is turbulent and is characterized by substantial mixing of water and contaminants.

Once the geologic setting is understood, the site hydrology must be evaluated. The location of the site within the regional ground-water flow system must also be determined.

Flow patterns (or nets) are in response to hydraulic gradients. In general, recharge areas are characterized by the downward flow of water from an area of high hydrostatic head to an area of low hydrostatic head (Figure 3). The directional flow of water in a discharge area can be downward, upward, or lateral. Recharge and discharge areas are either local or regional as shown in Figure 3. Note that in a given location, water in the upper portion of the saturated zone may move one way due to local hydraulic gradients and the opposite way in a deeper portion of the saturated zone where hydraulic gradients are different.

Knowledge of the hydraulically upgradient and downgradient direction from the site of interest is essential to proper development of a monitoring program. A graphical means of identifying these flow gradients is through flownets. Flownets are

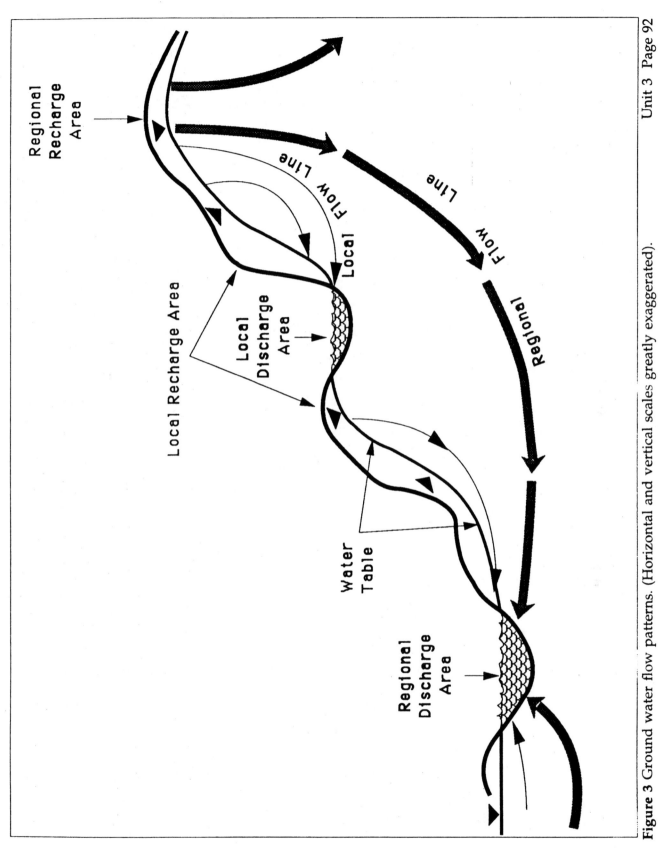

Figure 3 Ground water flow patterns. (Horizontal and vertical scales greatly exaggerated).

a two-dimensional model of the ground water system which illustrate the equipotential lines and flow lines. The flow lines are illustrated in a very elementary level with the flow paths shown within Figure 3. Equipotential lines are the location of equal hydraulic heads within the ground water system. The flow lines are oriented perpendicular to the equipotential lines. These equipotential lines are obtained through proper interpretation of ground water levels within wells or piezometers.

3.3.3 Aquifer and Other Terms

A variety of hydrologic terms are used when monitoring ground water. Although there is no universal agreement on how to define all of the terms, the commonly used definitions follow.

- aquifer--traditionally defined as a water-saturated geologic formation capable of yielding a significant (or economically important) amount of water to a well or spring. However, in ground water monitoring this term is used in a much broader sense to include all types of water-saturated zones. An aquifer is either confined or unconfined as follows:

  - confined aquifer--an aquifer that is overlain by geologic material that has a significantly lower hydraulic conductivity than the aquifer and is under pressure.

  - unconfined aquifer--an aquifer where there are no confining (or low hydraulic conductivity) beds between the saturated zone and the ground surface.

- aquitard--a water-saturated geologic formation that yields limited amounts of water to a well or spring.

- aquiclude--a water-saturated geologic formation that has low permeability and does not yield water freely to a well or spring, but may transmit appreciable water to or from adjacent aquifers. It is less permeable than an aquitard.

- permeability--a general term that expresses the relative ease with which a porous material can transmit a liquid.

- hydraulic conductivity--a numerical unit that describes the rate at which a fluid can move through permeable material.

- ground water--a general term applied to the water-saturated zone beneath the ground surface.

- water table--the upper most surface of the ground water or the undulating surface at which pore water pressure is equal to atmospheric pressure.

- vadose zone--the unsaturated zone consisting of soil, subsoil, water, and other materials extending downward to the water table.

- perched water table--unconfined ground water that is separated from the main ground water system by unsaturated and often low permeable (such as clay) material.

3.4 GROUND WATER MONITORING WELLS

3.4.1 Number and Placement

The number and location of monitoring wells should be based on a holistic approach that incorporates all site characteristics plus those of the adjacent areas, and carefully considers the project objectives. Yet the basic question remains of how much is enough. With respect to RCRA regulated hazardous waste facilities, the regulations help answer the questions of numbers and placement. The regulations (40 CFR 264) state that the ground water monitoring system must consist of a sufficient number of wells to yield samples from the uppermost aquifer that are (1) representative of the quality of background (or up-gradient) ground water that has not been affected by possible discharge from the facility and (2) representative of the quality of ground water down-gradient of the facility that might be affected by possible discharges.

Unfortunately, the exact meaning of these regulations in terms of the number and placement of monitoring wells is unclear. Ground

NOTES

water monitoring regulations (40 CFR 265) for interim status hazardous waste disposal facilities give more definitive information as follows:

- "Monitoring wells (at least one) installed hydraulically up-gradient from the limit of the waste management area. Their number, locations, and depths must be sufficient to yield ground water samples that are:

  — representative of background ground water quality in the uppermost aquifer near the facility; and

  — not affected by the facility."

- "Monitoring wells (at least three) installed hydraulically down-gradient at the limit of the waste management area. Their number, locations, and depth must ensure that they immediately detect any statistically significant amounts of hazardous waste or hazardous waste constituents that migrate from the waste management area to the uppermost aquifer."

In selecting the number and location of monitoring wells, a thorough understanding of flownets for the uppermost aquifer is necessary. As previously shown in Figure 3, flownets will tell the direction of ground water and product flow, both vertically and horizontally. This information, plus size of the facility, hydrogeology, statistical criteria for data analysis, and other factors will determine the location and the number of wells. For a simple ground water system, these steps may be as follows:

- determine aquifer geometry, thickness, and hydraulic conductivity (both vertical and horizontal);

- determine flow nets; and

- prepare a conceptual hydrogeologic model and plot monitoring zones.

In some situations, local characteristics can alter the flow of ground water. For example, when water is withdrawn from a well, a cone of depression forms in response to ground water removal.

Dimensions of the cone are largely dependent on the rate of water transmission within the aquifer and the rate of water removal; as the pumping rate increases and the transmission rate decreases, the cone of depression increases. The flow nets within the cone of depression are toward the well and in response to hydraulic gradients (Figure 4). Therefore, a down-gradient product plume can be drawn upgradient if the product is within the cone of depression.

3.4.2 Product Density and Solubility

The density of released product, in comparison to water, and its solubility in water are important in plume formation and, therefore, must be considered when designing a monitoring system. Figure 5 illustrates some of the simpler density-solubility relationships in plume formation.

For contaminants that have both soluble and insoluble components and a density greater than water, flow will be in two directions as shown in Figure 5A. Relatively insoluble contaminants with a density greater than water will respond to gravity more than they respond to the flow of water. Therefore, these dense, non-aqueous phase liquids (DNAPLs) can move against the flow of ground water in response to the greater force of gravity. Soluble components will be dissolved in the ground water and, once dissolved, move with the flow of ground water.

In contrast, a product with a density less than water will move with the flow of ground water (Figure 5B). If the product contains soluble components, these components will dissolve into the ground water and be carried with the flow. Low density insoluble compounds often termed light, non-aqueous phase liquids (LNAPL) will "float" at the ground water surface. Note that low density products can depress the water table level beneath the release. Like all floating objects, the product displaces water in proportion to its weight.

3.4.3 Selection of the Monitoring Zone

The placement of well screens determines the zone from which ground water is collected. A well screen is a section of slotted casing that allows for water entry. Following are general rules for selecting monitoring zones (or screen placement) as based on solubility and density of the released product.

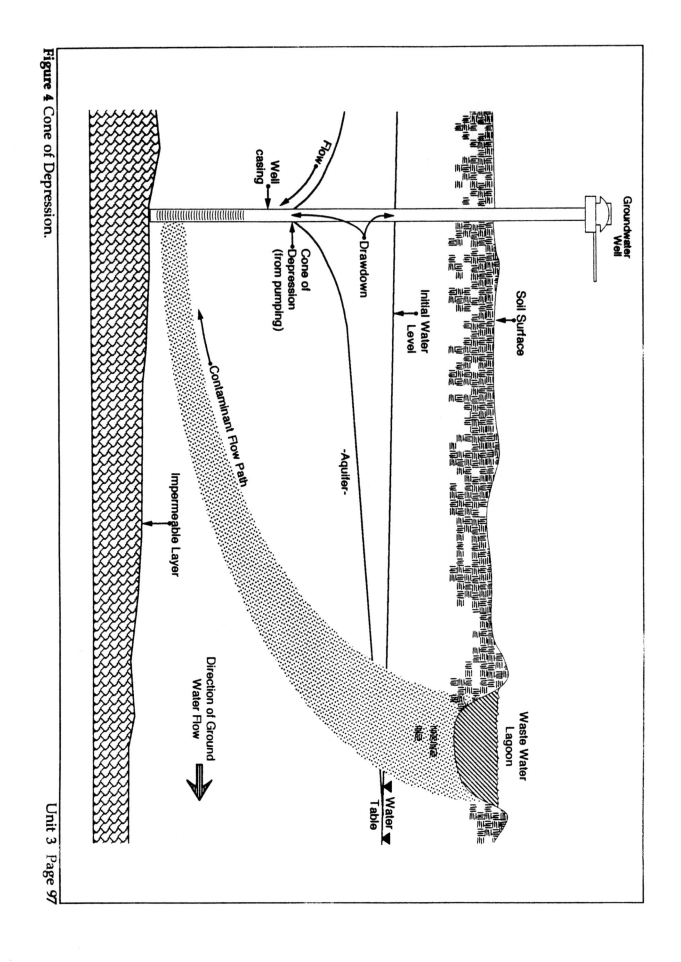

Figure 4 Cone of Depression.

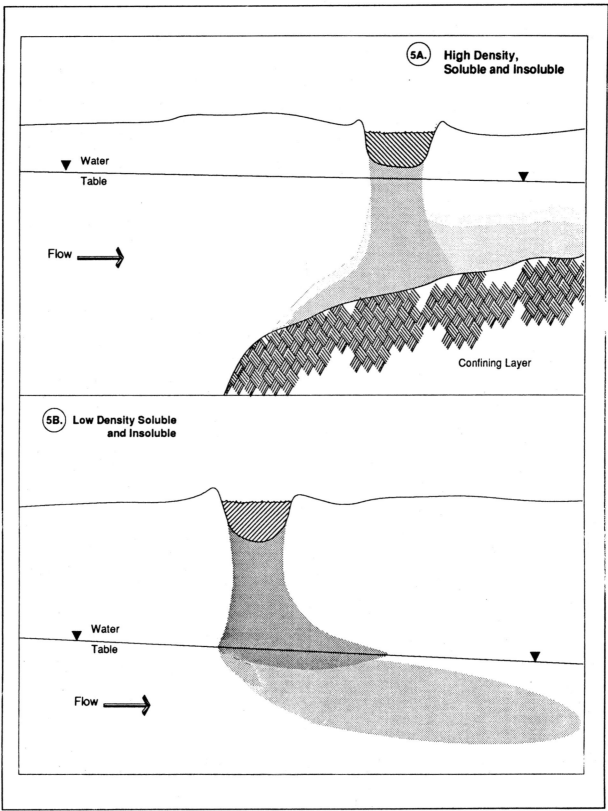

Figure 5 Density-solubility relationships in plume formation.

- For water-insoluble chemicals that float on water (specific gravity of less than 1.0), or for dissolved chemicals with a density lower than ground water, the screening interval should be the top of the aquifer. However, the screening interval must be long enough to accommodate any rise and fall of the water table.

- For chemicals that dissolve in water and the resulting water-product mixture that has a density similar to ground water, the screening interval is often located within the most permeable aquifer zone.

- For water-insoluble chemicals that sink in water (specific gravity of greater than 1.0), the screening interval is at the base of the aquifer. This also includes dissolved chemicals that have a higher density than ground water.

- For chemicals that are only slightly soluble, the screening interval may be the entire thickness of the aquifer.

These monitoring zones, as based on chemical characteristics, are illustrated in Figure 6.

3.4.4 Drilling

Drilling ground water wells uses the same basic technology employed in conventional geotechnical exploration. However, there are some significant differences, most relating to contamination and the ability to obtain discrete substrate samples. Drilling in contaminated areas will generate contaminated cuttings that may require disposal and contaminated equipment that must be thoroughly cleaned. Adequate precautions must be taken to protect workers from contaminants. Drilling and subsequent well instillation must be conducted to generate ground water samples (and often substrate samples) that are representative of existing conditions.

Following is a brief description of the common methods used in drilling ground water wells.

3.4.4.1 Drilling Without Circulating Fluids

When wells are to be installed in the uppermost aquifer and the substrate is composed of unconsolidated or loosely consolidated material, the drilling method does not require circulating fluids. Common drilling options are as follows:

Cable Tool Cable tool drilling is one of the oldest drilling methods still in use. These machines operate by repeatedly lifting and dropping a heavy string of drilling tools into the borehole. A drill bit breaks or crushes rock into small fragments and the reciprocating action of the bit mixes the crushes rock with water to form a slurry that is removed by a sand pump or bailer. This method is adaptable for most subsurface conditions, and the equipment is simple, maneuverable, and relatively quiet. The only waste generated is cuttings. Other advantages include minimal subsurface disturbance and contamination, low energy requirements, and suitability for drilling in boulder deposits and formations that are broken, fissured, or cavernous. However, cable tool rigs are slow. A drilling rate of 15 to 30 feet per day is typical in alluvium, and undisturbed substrate samples cannot be obtained. For these reasons, cable tool rigs are not commonly used for drilling monitoring wells.

Auger Drilling This is the most common method of drilling ground water monitoring wells. The method is fast and allows for the collection of "undisturbed" samples (Figure 8). A "continuous-flight" auger is commonly used. This consists of a steel shaft around which a continuous steel strip is welded in the form of a helix. Individual augers, called a "flight" and generally 5 feet long, are connected to one another such that the helix is continuous across the connections and throughout the depth of the borehole. A drill head is attached to the auger tip. The head consists of hardened or tungsten carbide steel inserted bits that serve as the cutting portion of the auger. The drill head is generally 10 percent greater in diameter than the diameter of the auger flights that follow. Augers generally rotate at 30 to 50 RPMs when the substrate is clay and up to 750 RPMs when the substrate is sand.

When substrate sampling is part of the well instillation process, samples can be scrapped from the helix when the auger is withdrawn. However, sampling depth is not accurate, and sample contamination is common because the outside portion of the sample is smeared along the inside of the drill hole when the auger is removed. To correct these limitations, a hollow stem

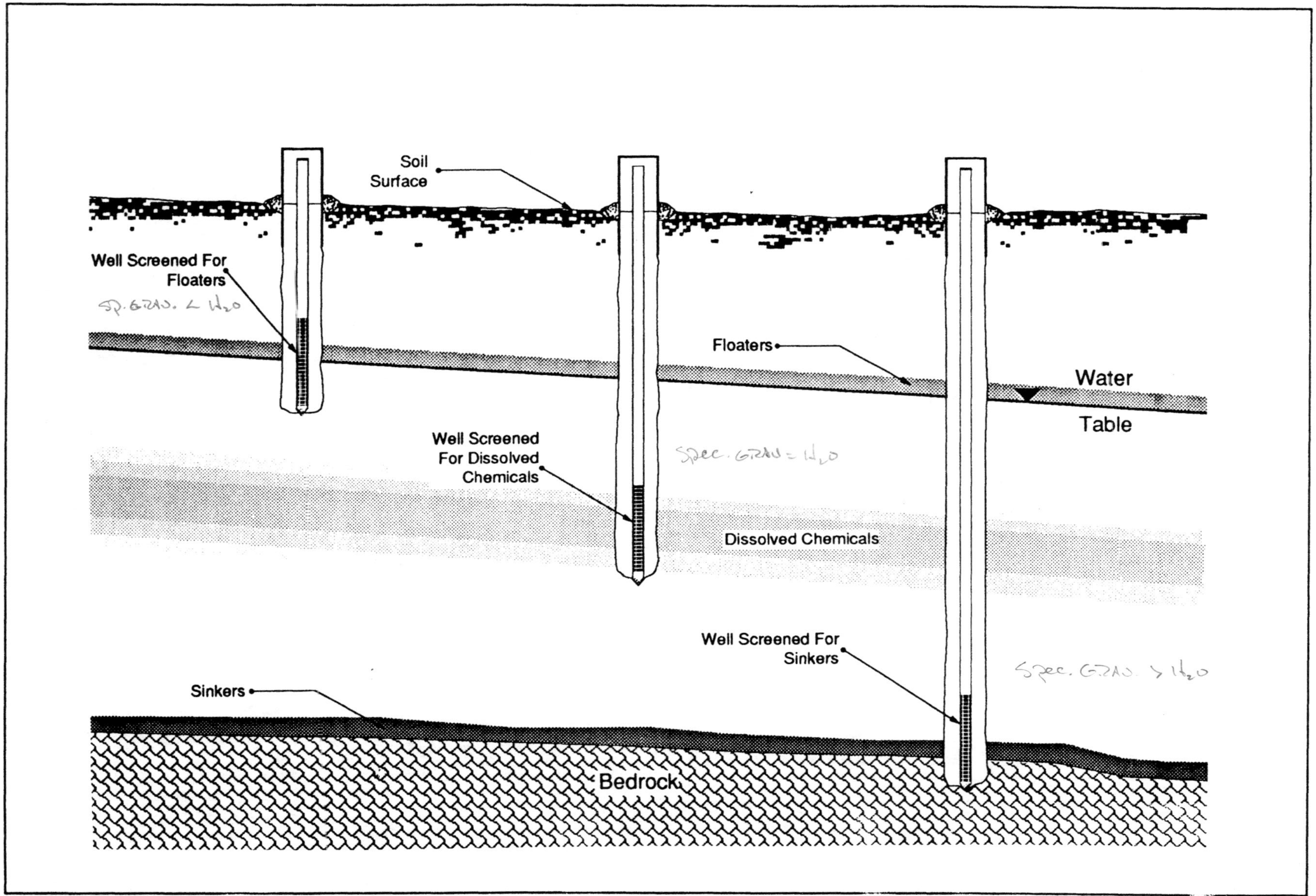

Figure 6 Monitoring zones for ground water wells.

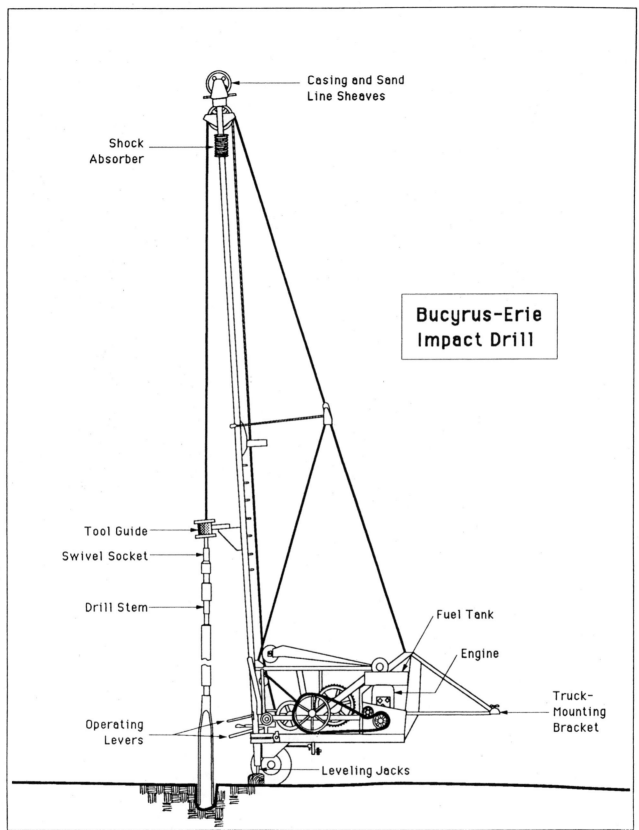

Figure 7 Cable tool drilling.

auger is commonly used. This auger is the same as that described above, except the steel shaft is hollow and the width of the helix is narrower (Figure 8). When augering without sampling, a center axil with a drill head is inserted to auger out the center portion of the hole. When the sampling depth is reached, auger rotation is stopped and the center axil is withdrawn. This leaves an open, cased hole into which a hollow sampling tube is inserted and lowered to the sampling depth. The sampling tube, generally 30 inches long, is driven into the undisturbed substrate below the auger. When the sampling tube is filled, it is withdrawn and the sample extruded. Most sampling tubes are of the "split spoon" type which is a sampling tube split lengthwise and held together with threaded rings on either end. To remove the sample, the top and bottom rings are removed and the sampling tube is opened like an unhinged clamshell.

Drilling machines using a continuous flight auger are of a top-drive design where all down force is applied directly to the top of the auger. The machines deliver a relatively high torque. Most machines are equipped with a hoist and a driving device for sampling.

The primary disadvantage to auger drilling is that this method cannot be used to drill rock, well-developed cemented layers, or boulders. Such conditions are commonly found in large alluvial basins in the Southwest and confined aquifers.

3.4.4.2 Drilling With Circulating Fluids

When auger drilling cannot be used due to consolidated substrate, drilling methods that use circulating fluids are used (Figure 9). The circulating fluid can be drilling mud (water with additives such as bentonite) or a gas (such as air). The circulating fluid is normally forced down through the drill pipe, out through the bit, and back up the annulus between the borehole wall and the drill pipe. The purpose of the fluid is to cool the bit, stabilize the borehole and remove the cuttings.

Rotary drilling is the most common drilling method when fluids are required. Here the drill pipe with an attached bit is continuously rotated against the face of the hole while circulating fluid is pumped through the system as described above.

Mud Rotary Drilling A circulating fluid consisting of water, clay minerals such as bentonite, and a variety of additives is used with

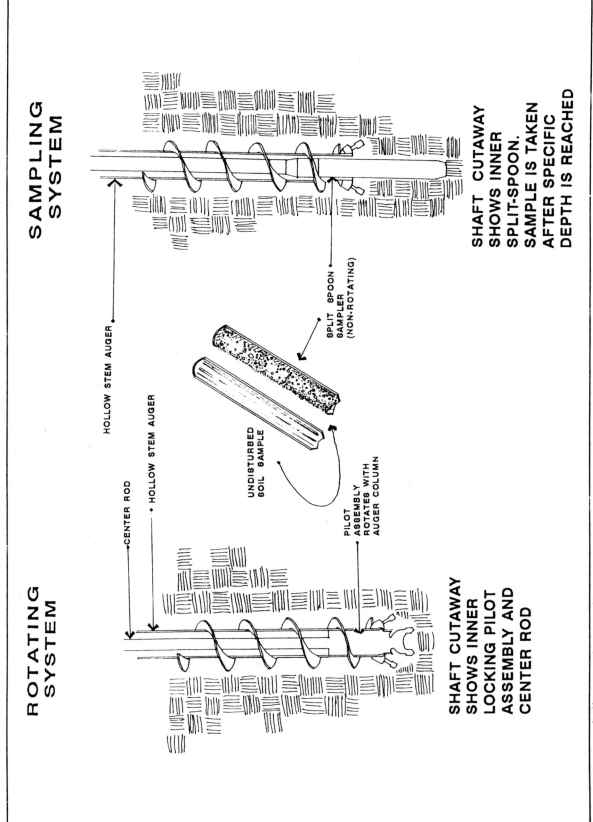

Figure 8 Auger drilling with sampling.

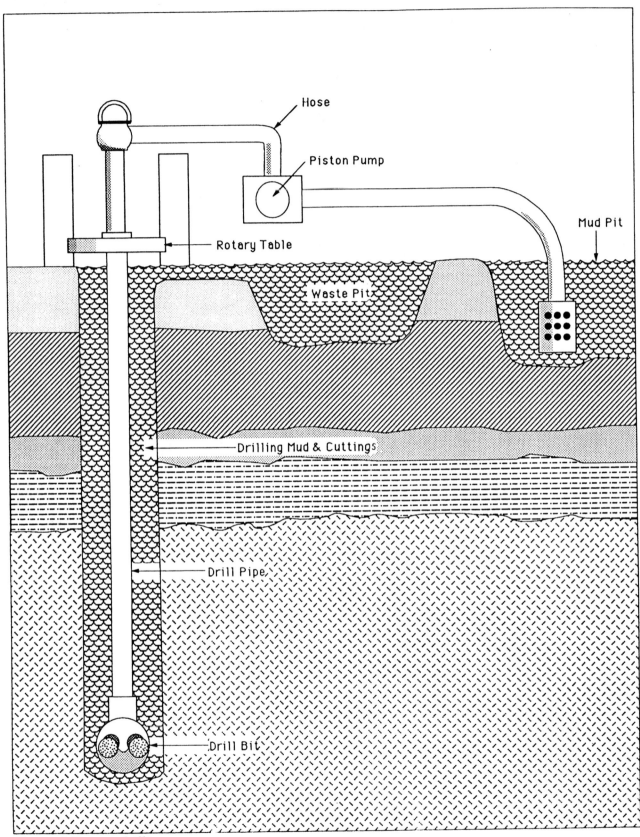

Figure 9 Rotary drilling.

mud rotary drilling. This method may not be the most desirable because (1) undisturbed samples cannot be easily obtained, (2) the drilling fluid may cross-contaminate the drill hole, (3) the drilling mud may invade the aquifer and affect the quality of subsequent water samples, and (4) large amounts of fluids are used that must be disposed of in an appropriate manner.

Air Rotary Drilling This method uses air as the circulating fluid. It avoids the contamination problems mentioned above and waste generation is usually lower. However, air rotary drilling can become messy when drilling through an aquifer because large volumes of contaminated water may be produced.

3.4.4.3 Selection of the Appropriate Drilling Method

Selection of an appropriate drilling method should be based primarily upon minimizing the disturbance of the geologic materials penetrated and the introduction of air, fluids, and muds. Also, the ability to penetrate the subsurface media must be considered. Other factors to be considered are discussed below.

Safety should be a major consideration. All drilling methods are hazardous because of the exchange of heavy drilling pipe, bits, sampling tubes, etc. and rotating equipment. In addition, air rotary drilling presents further personnel risks when drilling through a contaminated aquifer. Air drilling in this situation may produce a large volume of contaminated water that is difficult to manage.

Environmental contamination must be considered. All methods require appropriate protective measures, such as covering the ground with plastic and containing all waste products, and thorough decontamination of field equipment. Once these methods have been employed, auger and cable tool drilling presents the least contamination potential and rotary drilling the greatest potential.

Access can be limited, especially in urban areas. In general, auger and cable tool drilling use the smallest, lightest, and most maneuverable equipment. In addition, mast height is usually less than other types of drilling methods.

Noise can be a problem in urban areas. Cable tool drilling generates the least amount of noise followed by augering. Due to compressors and pumps, rotary drilling is noisy.

Waste generation must be considered when selecting a drilling method. Rotary methods produce the greatest amount of waste and the cable tool the least.

Depth of drilling can be a selection factor. Auger drilling is usually limited to a depth a approximately 150 feet, although augers have been used to drill holes as deep as 400 feet under favorable conditions. Rotary equipment can drill to depths of several thousand feet.

Sampling is an important consideration. When undisturbed samples are to be collected from unconsolidated material, the hollow-stem auger is the best method. When sampling consolidated material, mud rotary drilling is usually preferred.

Cost is usually not a selection criterion for drilling method. An inexpensive method often turns out to be the most expensive method because of well contamination, poor penetration rate, excessive time requirements, etc.

3.4.5 Permitting

In most states, a permit is required prior to drilling a monitoring well. In Colorado, the permit requires the name of the property owner, a legal description, the location of the well, the purpose of the well, the estimated depth, casing materials, the name of the driller, and other information. Once the well is drilled, the driller usually submits a drilling log and other information concerning the well. Finally, the permit is returned to the owner with conditions and a well permit number (see Appendix).

3.4.6 Well Design and Instillation

Figure 10 presents the design of a typical ground water well. Major components are well diameter, well casing, well screening, filter packs, annular seals, and surface protection.

3.4.6.1 Well Diameter

In most cases, monitoring wells are relatively narrow, typically 2 inches in diameter. Advantages of the 2 inch diameter will include the following:

- ♦ This diameter can accommodate the downhole equipment typically used in well monitoring.

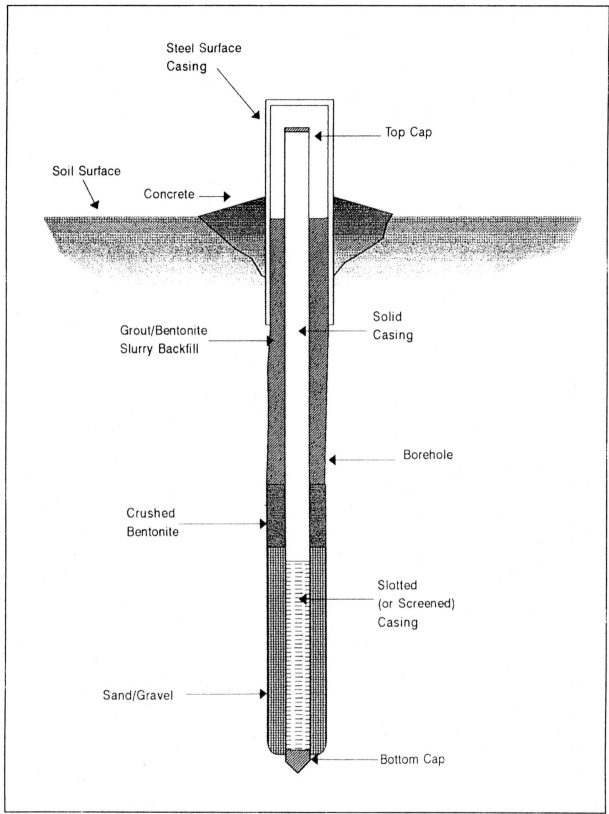

Figure 10 Cross-section of a typical ground water well.

NOTES

- The well is easy to develop because the smaller diameter requires less time due to the reduced volume of water involved.

- The purge volume of water is low, thus reducing sampling time and the cost of product disposal.

- The recovery rate is relatively fast, thus simplifying sampling.

- The cost is substantially less than larger diameter wells, usually two to ten times cheaper than the cost of a 4 or 6 inch diameter well.

3.4.6.2 Well Casing

The purposes of well casing are to prevent the collapse of geologic material into the borehole and to provide access to the ground water. To achieve these purposes, the casing material (including the screened section) must have adequate physical strength, chemical resistance, and chemical interference potential to meet site-specific conditions.

Strength characteristics include tensile, compressive, and collapse. Tensile strength is defined as the load required to pull the casing apart and is the most significant strength-related property for casing. As a general rule, the casing material should have enough tensile strength to support its own weight when suspended from the surface in an air-filled borehole. Compressive column strength is defined as the load required to deform the material by compressing it. Collapse strength is the ability of the casing material to resist collapse caused by any external loads; this property is proportional to the cube of the wall thickness.

Chemical considerations include corrosion of metallic casing materials and degradation of plastic materials. These processes will reduce casing strength and possibly alter the water chemistry. In addition, casing materials cannot assimilate chemicals from the ground water by adsorption onto the surface of the casing or by absorption into the matrix of the casing. The reverse is also important--casing materials must not desorb or leach contaminants into the ground water.

Following is a general discussion of the properties of the most commonly used monitoring well casing materials

Thermoplastics The most commonly used thermoplastic casing material is polyvinyl chloride (PVC). This material is resistant to corrosion, resistant to most chemicals and biological agents, has a high strength-to-weight ratio, is light weight, is durable, is low maintenance, has partial flexibility, and is low cost. Because of these characteristics, it is probably the most commonly used casing material. However, in comparison to metallic materials, tensile, compressive, and collapse strengths are low. Some chemicals will degrade PVC, particularly ketones, aldehydes, amines, chlorinated alkenes, and alkanes. This material may absorb some chemicals such as tetrachloroethane, bromoform, hexachloroethane, and tetrachloroethylene.

Fluoropolymers These materials consist of plastic monomers of which Teflon is the best known. These materials are highly resistant to chemical attack, even by highly concentrated and strong acids and organic solvents. They are also resistant to weathering, oxidation, and biological attack. Strength characteristics are a limitation; these materials have the lowest strength values of the commonly used casing materials. In addition, these materials are extremely flexible, and in some situations the casing becomes bowed when a load is placed on it. This can complicate or preclude sampling. Fluoropolymers can absorb certain chemicals such as tricholoethane, hexachloroethane, and tetrachloroethylene. Finally, these materials do not bond well with cement grout, and surface water could channel down the outside of the casing and into the ground water.

Metals Metals used for well casing include carbon steel, low carbon steel, galvanized steel, and stainless steel. These materials have superior strength over plastics. However, metal casing is prone to corrosion that can affect both strength and water quality. Corrosive conditions generally include low or high pH, greater than 2 ppm dissolved oxygen, presence of hydrogen sulfide, greater than 1000 ppm total dissolved solids, more than 50 ppm carbon dioxide, and more than 500 ppm chloride. The presence of corrosive conditions greatly increases the potential for contamination of ground water samples. For this reason, stainless steel is the only practical metal casing from ground water monitoring. However, even this metal can corrode in certain environments and contaminate the water with nickel and chromium.

Fiberglass Fiberglass-reinforced epoxy casing is nearly as strong as stainless steel but weighs about the same as PVC. The material is relatively inert in most environments, although some adsorption of volatile organic compounds can occur. Despite these favorable characteristics, this material is not in widespread use for monitoring wells.

Coupling procedures for joining casing segments must also be addressed. Plastic can be joined by solvent cementing, but this is rarely used because the solvent can contaminate ground water samples. Flush-joint threaded casing is the most popular. Most designs use square threads that are easier to screw together and less likely to become cross threaded than the V-shaped threads. Some flush-joint coupling designs include O-rings to ensure effective coupling; the rings are usually made of Viton or nitrile. Stainless steel can be welded, but this is difficult and time consuming; threaded joints are commonly used. However, the joints are usually made of a short section of pipe welded inside of the casing. This creates a rough joint that effectively reduces the diameter of the well.

3.4.6.3 Screening

The purpose of a well screen is to provide open areas that permit water to flow from the formation into the well. The amount of open area in a well screen is a critical decision in well installation and is largely dependent on parameters of the filter pack as described in the next section.

Only commercially manufactured well screens should be used in monitoring wells; the slotted well screen and the continuous-slotted screen are commonly used (Figure 11). The slotted well screen is made from standard well casing into which horizontal slots of predetermined widths are cut at regular vertical spacing by machining tools. Slot openings are designated by numbers corresponding to the width of the opening in thousandths of an inch. For example, a number 10 slot has an opening of 0.010 inches. The continuous-slotted screen is manufactured by winding wire or plastic, often triangular in cross section, spirally around a circular array of longitudinally arranged rods. At each point where the wire or plastic crosses, the two pieces are securely joined by welding to create a one-piece rigid unit. The slots are created by spacing the successive turns of wire or plastic at the desired intervals.

The choice between slotted screen and continuous-slotted screen must consider several parameters. For a screen of the same slot

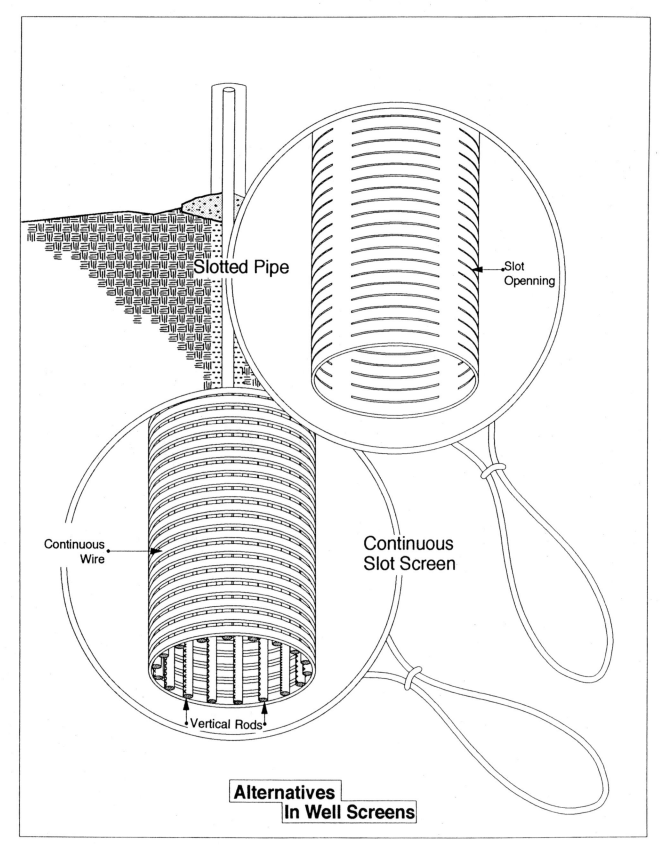

Figure 11 Screen types.

size, the continuous-slotted screen has at least twice the open area of the slotted screen. The greater open area for the continuous-slotted screen allows water to enter the well at a faster rate, thus increasing the efficiency of well development and purging. In addition, triangular-shaped wire is often used for the continuous-slotted screen. This wire shape allows particles slightly smaller than the slots to pass into the well without wedging in the slots and eventually clogging the screen. Unfortunately, continuous-slotted screen is usually substantially more expensive than slotted screen.

Screening length should be as short as possible to achieve monitoring objectives. Screening length is usually between 2 and 10 feet, and rarely exceeds 20 feet. Shorter screens provide more specific information about water quality than longer screens. However, if the surface of the water table is being monitored, the screening length should be long enough to accommodate seasonal fluctuations in the water table.

As previously discussed, the interval of the ground water zone to be monitored must be selected with care to ensure that appropriate samples are collected. If chemicals with a specific gravity of less than 1.0 are to be monitored, the screen interval must be at the top of the saturated zone. If chemicals with a specific gravity of more than 1.0 are to be monitored, then the screen intervals must be placed at the base of the saturated zone. For dissolved chemicals, the screened interval is commonly in the middle of the saturated zone. If integrated samples are to be collected (floaters, sinkers, and dissolved chemicals), then the full length of the saturated zone should be screened.

3.4.6.4 Filter Packs

The purpose of the filter pack surrounding the screened portion of the ground water well is to provide a zone of high hydraulic conductivity and to prevent fine-grained formation materials (clay, silt, or fine sand) from entering the well.

In some cases, the natural material surrounding the well will collapse around the well screen and form a natural filter pack. This may work in areas where the natural formation material is coarse-grained, permeable, and relatively uniform in grain size. The primary disadvantage associated with using the natural formation is longer well-development time.

Artificial filter packs are often used and consist of placing a coarse to medium sand size material (sand-pack) into the annular space between the well screen and the borehouse wall. Factors that should be considered when designing a filter pack include (1) grain size and well screen slot size, (2) grain size distribution, (3) grain shape properties, (4) filter pack dimensions, and (5) filter pack material type.

The primary function of the filter pack is to trap small particles that will flow into the well with the water. Therefore, the filter pack grain size should be based on the grain size of the finest formation material to be screened. Once this decision is made, the size of well screen slots can be made. As a general rule, the well screen should allow less than 10 percent of the filter pack to enter the well. The filter pack material that does enter the well is removed during well development. However, the filter pack grain size can be too small. If this is the case, the filter pack may be lost into the formation and the voids left by the lost material filled by formation material that allows fine particles to enter the well. Filter pack grain size is one of the most important decisions in well design.

With respect to filter pack grain size distribution, uniform grain size is preferred to graded sizes for monitoring wells. The filter pack should be composed of uniformly sized particles because the slots in the well screen are of uniform size. In addition, non-uniform filter packs will segregate during placement, thus creating areas of coarse material and areas of fine material around the screen.

Filter pack grain shape considers the roundness and sphericity of the individual grains. Particles that are less round and less spherical tumble and oscillate as they fall through water during placement. This slows the rate at which these particles fall and increases the likelihood of segregation of the particles and the potential for bridging which creates a void within the filter pack.

Length of the filter pack should extend from the bottom of the well screen to at least 3 to 5 feet above the top of the screen. This allows for settling of the filter pack and gives a buffer for the grout that is applied to the top of the filter pack. In addition, the grout will penetrate the upper portion of the filter pack to some extent, so an extension of the filter pack is necessary to prevent grouting any of the screened section. Width of the filter pack should be thick enough to completely surround the well screen, but thin

NOTES

enough to minimize the resistance caused by the filter pack to the flow of water into the well. In practice, a filter pack thickness of 3 to 5 inches is most common.

Filter pack material should be either quartz sand or glass beads. This material should be washed, sized, dried, and packaged at the factory. In no case should the filter pack be made of crushed stone.

Methods of filter pack installation include gravity placement, placement by tremie pipe, reverse circulation placement, and backwashing.

Gravity placement is the most common, but this method should be limited to relatively shallow wells with a annular space greater than 2 inches. This will minimize bridging and segregation of the filter pack during placement. However, some segregation will occur because as the sand falls through the water column, the finer grains fall at a slower rate than the coarser grains. Therefore, coarse material dominates the lower portion of the filter pack and finer materials comprise the upper portion. Additionally, gravity placement maximizes the potential for formation materials along the borehole wall to mix with the filter pack, thus reducing its effectiveness.

The tremie pipe method uses a pipe to direct the filter pack to the interval being screened. Initially, the pipe is placed at the base of the screened interval and filter pack material is poured down the pipe. When the filter pack has filled the annular space around the intake, the pipe is pulled up to a new position and the process repeated. This minimizes both filter pack segregation and incorporation of borehole wall material. This method can be used with hollow-stem augering. Here the hollow-stem auger serves as the tremie pipe. The auger is withdrawn 1 to 2 feet at a time while filter pack material is added.

Regardless of the application method, the filter pack should be applied slowly to minimize bridging. In addition, vibrating the casing either by hand or with a mechanical vibrator is often done to minimize bridging.

The reverse circulation method uses a sand and water mixture that is fed into the annulus around the well screen. The water flows into the well and is then pumped up through the casing, leaving the filter pack behind. However, this method introduces water into the borehole, and the potential for monitoring well contamination severely restricts the use of this method.

The backwashing method allows the filter pack material to free-fall down the annulus while concurrently injecting water into the well casing, through the well screen, and back up the annulus. Due to the potential for well contamination from the injected water, this method is rarely used.

3.4.6.5 Annular Seals

The annular space from the top of the filter pack to the surface must be completely sealed to provide protection against surface water and contaminants from infiltrating into the well. In addition, sealing the annular space increases the life of the casing by protecting it against degradation and providing additional structural integrity for the casing. The most common types of annular seal materials are bentonite and Portland cement.

When mixed with water, bentonite can expand its original volume 8 to 10 times to form a dense clay mass that produces a tight seal between the well casing and the adjacent formation materials. This material is available in chips and pellets. It is generally prepared by mixing bentonite into water at a ratio of approximately 15 pounds of bentonite to 7 gallons of water; yield is one cubic foot of grout. The grout is then pumped through a tremie pipe to its intended position in the annulus. However, when a tremie pipe is used, the grout may plug the pipe. Often, the grout is poured down an open hole, or the bentonite chips or pellets poured down the annular space and allowed to partially hydrate.

The ability of bentonite material to form a tight seal depends upon additional expansion after emplacement. This expansion will require water; therefore, the geologic formation in which a bentonite grout is emplaced must supply some water to complete hydration and maintain the bentonite in a hydrated state. Because of its requirement for hydration, bentonite is generally placed to form a layer 1 to 5 feet thick above the filter pack; bentonite rarely extends to the surface.

Portland cement mixed with water is another type of commonly used grouting material. It is generally mixed in the proportion of 5 to 6 gallons of water to one bag (95 pounds) of cement and placed by using a tremie pipe. However, a cement grout will pull away from either the casing or borehole wall as it dries, thus destroying the integrity of the seal. Additives must be added to the cement to compensate for the shrinkage. Common expansion additives and their percent by volume in the cement mixture are as follows:

>bentonite (3 to 5 percent);
>gypsum (3 to 6 percent); or
>aluminum powder (1 percent).

Bentonite and particularly cement can alter ground water chemistry by increasing the pH and/or introducing trace elements. When this is of concern, the contamination potential can be minimized by placing a very fine-grained secondary filter pack on top of the primary filter pack.

In practice, both bentonite and cement are commonly used as grouting agents to seal the annular space. A bentonite plug, 1 to 3 feet long, is often placed immediately above the filter pack. The remaining portion of the annulus is grouted with a cement-bentonite slurry.

3.4.6.6 Surface Protection

Completion of a monitoring well includes surface protection to prevent surface runoff from entering and moving down the annulus of the well and to protect the well from damage. Protection generally consists of a surface seal and a protective casing.

The surface seal is often an extension of the annular seal previously discussed. The annular grout is extended to provide an apron radiating 2 to 3 feet from the well and tapered away from the well to discourage runoff from entering the well area.

The protective casing is either an above ground type or a flush-to-ground type (Figure 12). For the above ground type, a protective casing or standpipe is set into the surface seal while the seal is still wet. Approximately one half of the standpipe is anchored into the cement and the other half extends above the seal to protect the well casing; the inside diameter of the standpipe should be large enough to allow easy access to the well casing. Standpipes are generally made of carbon steel. As with the well casing, the standpipe should be vented to prevent the accumulation of vapors. Additionally, the standpipe should have a drain hole just above the top of the apron to allow the drainage of any water

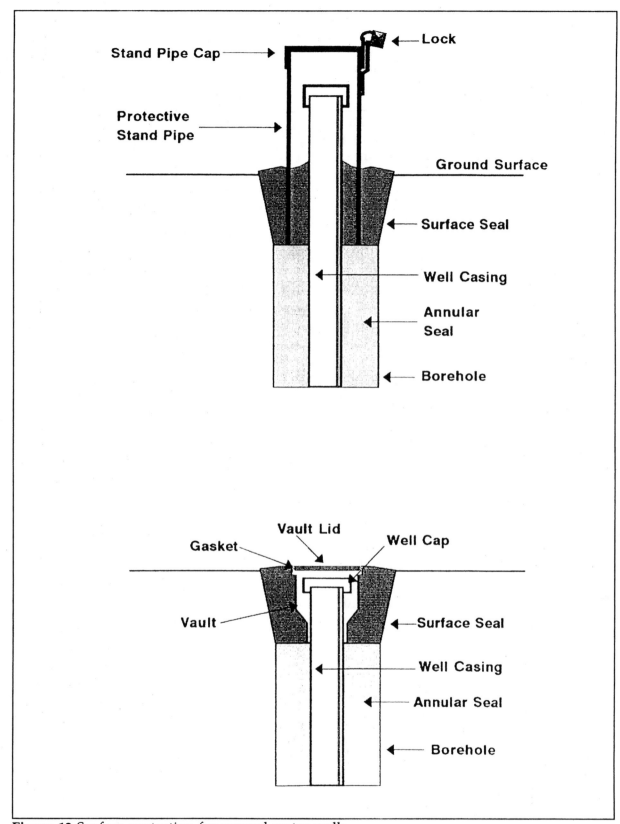

Figure 12 Surface protection for ground water wells.

accumulating within the pipe.

A flush-to-ground protective casing is similar to the above ground type except the casing (often a utility vault of some kind) does not extend above ground level. The vault lid is usually fitted with a flexible O-ring to provide a seal because of the increased potential for surface water entry.

Finally, completion usually requires some type of lock to limit access to the well and labeling to identify the well.

3.4.7 Temporary Ground Water Wells

A variety of techniques are used to install wells designed for one-time sampling or sampling over a short period of time. Wells can be drilled as described above, but once the screen and casing are placed, a filter pack and grout annular seal may not be installed. When the sample has been collected, the casing is removed and the well is backfilled with grout. If the well is to be monitored for a short period of time, a filter pack is installed and a bentonite seal placed above the filter pack. Grout is often not placed in the remaining annular space.

Another technique is called the "Hydropunch." Here a hollow-stem auger is used to drill to a depth just above the sampling zone, generally about a foot above the top of the ground water zone. A drive cone and hollow drive casing are inserted into the hollow stem auger and driven into the underlying material. The drive casing is then partially withdrawn, exposing a well screen. A sample of the ground water is collected as described in the next section. The drive casing and associated materials are removed and the well closed. This system has the advantages of low costs, minimal disturbance of the sampling zone, and the ability to collect samples that have not been averaged or diluted by well development and purging. Other systems are being developed for sampling shallow aquifers. For example, one promising system hydraulically pushes a probe into the ground without drilling.

3.4.8 Decontamination

Drilling equipment can be decontaminated using a high pressure, hot water wash system. This system commonly consists of an initial wash with soapy water and scrub brushes. This is followed by a high power clean water rinse to remove soap and contaminants. The equipment is then allowed to dry before use.

This system is often incorrectly referred to as "steam cleaning," perhaps due to the fine, steam-like mist that is emitted from the high pressure spray.

The second method uses high pressure steam to remove contaminants. This system minimizes the volume of waste water, but has the disadvantages of not being able to effectively remove large particles and clayey soils, or the logistics of generating sufficient steam under field conditions.

Care must be taken not to spread drill cuttings and other potentially contaminated material around the site. Cuttings and protective equipment are usually placed in drums, labeled, and left on-site, or disposed of appropriately. Specific procedures may vary according to contaminant and regulatory requirements.

3.5 WELL DEVELOPMENT

A well cannot be successfully sampled until it has been "developed." The purpose of this development is to remove drilling contamination and to repair the subsurface damage associated with drilling. Contamination is often in the form of fined-grained material within the sand pack surrounding the screen and from the surrounding soil or geologic formation. Drilling fluids, particles of casing material, and associated materials may likewise contaminate a new well. Drilling disturbs the natural water-bearing formation due to compaction and invasion of fine-grained particles into the water-bearing formation. The objective of well development is to remove the contaminants and to restore the original permeability.

All well development methods force water through the well casing, sand pack, and into the formation. Once this surging or flushing is completed, the well is pumped or bailed to remove the debris and contaminants.

Although full development of a well prior to sampling is necessary, overdeveloping should be avoided because this can cause compaction of the filter pack or loss of the filter pack into the formation.

3.5.1 Flushing and Surging

Following are the common techniques for well flushing and surging.

3.5.1.1 Overpumping

Probably the simplest method of well development is pumping at a higher pumping rate than will be later used during well sampling. If a high pumping rate will flush the sediment, the low pumping rate used during sampling should not generate significant sediments.

Electric submersible pumps are commonly used for overpumping. In some cases, electric "well development" pumps that can remove water at a rapid rate will be used for this operation. These pumps may not be compatible with sampling objectives due to the pump action and materials used in pump construction; therefore, such pumps are limited to well development.

This technique is limited in that most of the development action takes place in the most permeable zone closest to the top of the screen. Therefore, water will move preferentially through this developed zone and leave the rest of the well poorly developed and unable to contribute significant water during sampling. Therefore, a representative sample may not be obtained. In addition, overpumping allows water to flow in only one direction, into the casing, during development. This can create bridging of the particles outside of the casing. These bridges eventually break and sediment enters the well. Dewatering the formation and water disposal are other potential problems.

One method to minimize the creation of sediment bridges and to obtain better cleaning of the filter pack is to alternate the pumping rate from fast to slow. This creates an in-and-out washing motion from the well, through the screen, and into the filter pack. When clear water is obtained during both cycles, the well is developed.

3.5.1.2 Backwashing

Backwashing can be used in conjunction with overpumping. The purpose of backwashing is to reverse the flow of water out through the screen and into the surrounding sand pack and formation. Sediment bridges are broken down by flow reversal and agitation. This sediment then moves into the well and is removed. Backwashing is usually accomplished by adding water to the well.

One disadvantage to backwashing is the addition of water to the well. All of this water may not be removed during development, thus affecting the analytical results from subsequent water sampling. In addition, full development of the well is unlikely when backwashing is used. Generally, only the most permeable zone of the formation is adequately developed.

Backwashing can also be accomplished by cycling a ground water pump to allow water to reverse flow back into the well from the discharge line. For shallow wells, this can be readily completed using a peristaltic pump with the discharge line placed in a water container and cycling the flow direction of the pump.

3.5.1.3 Mechanical Surging

This method forces water into and out of the well by raising and lowering a plunger-like device. This plunger could be a bailer, but a more effective tool is a metal (usually stainless steel) cylinder with a rubber or leather flapper the same diameter as the inside diameter of the well. When using this method, lower the surge until it is below the water table and start a gentle up and down surging action. Once the water begins to move easily into and out of the well, the surge is lowered farther into the well and the surging continued. The deeper the surge is in the well the greater the force of the surging movement.

As with the other two methods mentioned, surging often develops the most permeable zone. In addition, for surging to be effective, the rubber or leather discs must properly fit the well casing. Caution must be used with surging to avoid damage to the well casing, especially at casing joints.

3.5.1.4 Air Surging

Air surging injects air into the well to lift the water column. When the water reaches the top of the casing, the air supply is cut off; the falling water column causes a surging action within the well. Key components of the air surging system are the air line and the surrounding eductor pipe. The eductor pipe is necessary when limited volumes of air are available or when the water level is low in relation to the well depth. Air surging can also function similarly to water jetting when special nozzles or perforated piping is used.

A limitation of air surging is the introduction of air into the well water. This may interfere with some water analyses, such as volatile organics. In addition, a considerable amount of equipment is necessary and water is often spilled around the well during surging.

3.5.1.5 Water Jetting

This consists of a horizontal water jet inside the well that produces a high velocity water stream that shoots out through the screen. This action agitates the sand pack and cleans the formation. Equipment consists of a water jet with two to four nozzles, a high pressure pump, pipe, and a water supply. The jet is placed near the bottom of the well and slowly rotated while being pulled up, thus exposing the entire length of the screen to the water jet action.

This method is highly effective in well development because the water is forced out of the screen and into the formation. However, water must be added to the well and this can cause sampling and analytical problems if all of this water is not removed. Also, pressure must be closely monitored to prevent damage to the casing material.

3.5.2 Removal of Debris and Contaminants

Following one of the above methods of flushing or surging, the well is pumped or bailed to remove the debris and contamination. In most cases, measurements are taken after the removal of each casing volume as described in a later section.

Following removal of a casing volume, pH, conductivity, turbidity, and other parameters are generally determined. Water removal continues until one or more of these parameters becomes constant. At this point it is assumed that the water entering the well is from the surrounding formation and that all significant contamination and debris have been removed from the well. An example of a well development form is presented in Figure 13.

3.5.3 Slug Testing

It is important to know the rate at which water passes through a well for several purposes. This includes knowing whether the well has been sufficiently developed, determination of sufficient well purging prior to sampling, and estimation of the spatial distribution of a contaminate. One key parameter which must be measured is hydraulic conductivity, and a slug test is frequently used to estimate this parameter. The slug test is appropriate where the hydraulic conductivity is moderate to low; formations

WELL DEVELOPMENT LOG

Well Number_____   Job Number_____

Date Started_____ Time Started_____

Date Completed_____ Time Completed_____

Field Personnel_____

Development Method_____

WELL INFORMATION

Description of Measuring point (MP)_____

Total   Depth of Well Below MP, ft_____

        Depth of Water below MP, ft_____

        Water Column in Well, ft_____

        Gallons in Well _____ (gallons/casing volume)

FIELD PARAMETERS

Time	Casing Volume	Cond.	Temp.	pH	Turbidity	Drawdown
____	____	____	____	____	____	____
____	____	____	____	____	____	____
____	____	____	____	____	____	____
____	____	____	____	____	____	____
____	____	____	____	____	____	____
____	____	____	____	____	____	____
____	____	____	____	____	____	____
____	____	____	____	____	____	____
____	____	____	____	____	____	____

Comments_____

_____

Note: for a 2 inch diameter well casing, there are 0.16 gallons per foot of water depth.

Figure 13 Well development form.

with a high conductivity will recover water too quickly for accurate measurements. However, the use of electronic data loggers can facilitate more accurate readings under conditions of rapid data acquisition. Areas where slug tests are inappropriate may use multiple-well tests whereby one well is pumped at a high rate such that it causes the water levels of nearby wells to become lowered.

If the well screen extends into the unsaturated zone above the water table, the slug withdrawal technique is appropriate. This method uses a float consisting of a long piece of weighted pipe with a diameter slightly less than the well casing and sealed at both ends. The float is lowered into the well until it reaches a level at which it floats. The presence of this float will force water up the well and above the static (or original) water level. The float remains in the well until the water head decreases to the original static level. At this time the float is quickly removed, thus instantaneously dropping the water level; the volume of water drop is equal to the volume displaced by the weighted float. A device is then inserted into the well to measure the water level (as a function of time) until the static water level is reestablished. These measurements are then converted into hydraulic conductivity. Instead of the float method described above, a high volume pump can be used to create an instantaneous withdrawal of water.

If the screened portion of the well does not extend into the unsaturated zone, the slug injection method can be used. This method quickly adds a known volume of water to the well and then measures the decrease in water column height as a function of time.

Alternatively, rapid application of a high pressure gas beneath the ground water surface will blast water out of the well when the static water level is not excessively deep. After this instantaneous drop in water level, water recovery is measured until the static water level is reestablished. However, this method has the major disadvantage of discharging potentially contaminated water out of the well.

A variety of methods exists for calculating the hydraulic conductivity once the necessary data inputs have been collected. Most of them are based upon Darcy's law: $V=Ki$ (V:specific discharge m/s; K:hydraulic conductivity m/s; i:hydraulic gradient). The necessary input parameters for slug tests are the changes in water height as a function of time and various well dimensional parameters.

3.5.4 Surveying

In most cases, it is necessary to locate the well by surveying. Well locations are then plotted on a map and the information used to develop and interpret hydrogeologic data. Well elevations are also necessary. Generally, elevation of the standpipe is determined and permanently marked on the standpipe. During sampling, this reference point is used to calculate water table elevation. Such data are needed to determine direction of ground water flow and other parameters.

3.6 GROUND WATER SAMPLING

Following development, the ground water well is ready for sampling. This section assumes that sampling objectives, containers, preservation, shipping, analysis, QA/QC, etc. have been documented as previously discussed.

3.6.1 Well Approach

As with all sampling efforts, personal safety is the first concern. Therefore, a ground water well must be approached with caution. Explosive gases can accumulate in the well and harmful vapors can surround the well.

Wells should be approached from upwind, thus allowing the wind to disperse any airborne contaminants away from the sampler. Appropriate air monitoring instruments should be used to detect suspected contaminants as the sampler approaches the well. Once at the well, air monitoring instruments are used to check for hazards around the well head, especially in the vicinity of the standpipe cap and vent hole. Next, carefully remove the standpipe cap and monitor the air inside the standpipe. Finally, remove the casing cap and monitor the air inside the well casing.

When sampling, do not stand directly over the well because contact with splashed water is possible. In addition, sampling may bring up toxic air in the well.

Following this monitoring, revise respiratory protection procedures as needed. Often, the next step is to check the ground water for light, non-aqueous phase liquids, or floaters. Interface probs or clear plastic bailers are often used.

3.6.2 Well Purging

Purging a well prior to sampling is usually required because the water standing in the well casing cannot be considered representative of formation water. However, considerable debate exists regarding the volume of water that must be removed and the most appropriate method to accomplish purging.

In comparison to formation water, the standing water in the well is potentially different due to:

- limited interaction with formation water and formation geology;

- contact with the well casing and filter pack material;

- contact with the atmosphere; and

- different chemical reactions.

As a result, water standing in a ground water well may have a different pH, Eh (oxidization-reduction potential), temperature, dissolved solids, volatile organics, dissolved gases, and conductivity than the formation water. The purpose of purging is to allow formation water to replace the standing well water while minimizing disturbance to the ground water flow regime.

The amount of water that must be removed to obtain a representative formation sample can be determined in several ways. The three most common techniques are briefly explained below.

3.6.2.1 Casing Volume Criterion

Federal, state, and company sampling protocols often state that purging can be accomplished by removing 3 to 5 casing volumes of water. A casing volume is usually defined as the volume of standing water in the well. However, some people define casing volume as both the water volume in the well and within the filter pack. For wells that have a slow recovery rate, it is suggested that the well be pumped dry and allowed to recover (within a defined time limit) before sampling. However, these rules of thumb do not apply in many cases. Research has shown that purging volumes are dependent upon many factors (e.g. analyte of interest, location of pumping device within well, hydraulic performance of well) such that the appropriate casing volume may range from zero to twenty.

3.6.2.2 Parameter Stabilization Criterion

This protocol specifies that purging is not complete until specified field parameters have stabilized to within some predetermined level of precision. The most common field parameters are pH (generally considered the most sensitive), temperature, conductivity, and Eh. Typically, these parameters are measured after each casing volume is removed or the measurements may be obtained as the well is being purged if in-line analytical sensors are used. When there is no significant change in field parameters for two consecutive casing volumes (or for some other specified volume), the well is considered purged. The applicable protocol will vary depending on regulatory requirements and site specific conditions.

3.6.2.3 Hydraulic & Parameter Stabilization Criterion

This requires the calculation of aquifer transmissivity (K times the saturated thickness). Knowing this parameter along with the casing diameter and purging rate, the length of purging time required to obtain a 100% formation water can be calculated. This calculated purge volume is then confirmed by following the parameter stabilization criterion. While this approach may have superior scientific merit than the other approaches, parameter stabilization should yield equivalent results.

3.6.2.4 Purging Rate

Purging should be at rates below that used to develop the well and below the recovery rate so that migration of water into the formation above the well screen does not occur. In addition, slow purging prevents dewatering the saturated zone. Slow purging rates also minimize ground water aeration, prevent the stripping of volatiles from the water, and reduce the likelihood of mobilizing colloids in the formation that are immobile under natural flow conditions. Ideally, purging rates should be less than approximately 0.2 to 0.3 liters per minute; in practice, purging rates are often much faster.

3.6.2.5 Purging Volume

While the purging volume must be sufficient to allow a representative sample to be collected, excessive purging also creates problems. Overpurging can lower the water table and consequently dewater a portion of the saturated zone. This exposes the dewatered zone to air and other gases and to substances floating on top of the water table. When the purging ends and the water level recovers, both liquid and gaseous contaminants can be mixed with the water and introduced into the sample. In some cases, overpurging can dilute contaminants to the point where they are no longer detectable. In other situations, overpurging can cause significant drawdown, thus drawing contaminants into the well that would otherwise flow away from the well. This not only contaminates the sample, but enlarges the contamination plume.

Although purging is common practice prior to ground water sampling, purging is not recommended in some situations. For example, purging may not be applicable when volatile organics are monitored in fine-grained sediments. The fine sediments can strip the volatile organic compounds from the well water. Purging can also cause pressure changes that force dissolved gases out of solution. This in turn causes other chemical changes, such as precipitation of calcium carbonate and heavy metals on the well screens. Purging of slow recovery wells introduces air into the formation. Studies have shown that up to 70 percent of the volatile organics in water can be lost once air enters the formation.

As indicated above, slow recovery wells often create purging and sampling problems. If purging is necessary for such wells, some recommend that sampling occur as soon as the original static water level is reestablished. Others recommend that sampling be conducted during well recovery. Common practice is to sample the well the day after purging.

3.6.2.6 Purging Location

That portion of the well to be purged prior to sampling must reflect sampling objectives if a representative sample is to be collected. For example, if the purging location is at the surface of the water table, water in the middle or bottom of the water column may be relatively unaffected and may remain non-representative of formation water. Therefore, purging depth should coincide with sampling depth to ensure that formation water is sampled.

3.6.2.7 Purge Reduction

Due to the purging problems mentioned above and the often large volumes of contaminated water generated during purging, reducing the volume of water removed during purging will decrease labor costs, reduce purge water disposal, and will allow the sampling program to collect samples that more accurately represent an undisturbed aquifer. Methods and equipment to minimize purging volumes (often called micro-purging) are briefly discussed. All methods use the parameter stabilization criterion to determine when purging is completed.

Low-flow sampling devices, such as bladder pumps and centrifugal pumps, can purge the water from the sampling zone without disturbing other stagnant water within the casing or mixing water from other vertical zones. In addition, low flow rates reduce turbidity, degassing, and contaminant dilution. Because the insertion of pumping devices into a well causes mixing within the water column, the low-flow pump is often dedicated to the well. However, low-flow purging is often ineffective when purging a zone of low hydraulic conductivity that is overlain with zones of high hydraulic conductivity.

Packers can be used that partially isolate the purging and sampling zone within a well. Hydraulic or pneumatic activated packers are lowered into the well and inflated when the desired depth is reached. When inflated, they wedge against the casing wall or screen and allow purging and sampling to take place within an isolated portion of the well (Figure 14). Packers are made of rubber or rubber compounds and can be used with most types of pumps. Using packers, a specified portion of the well is isolated, purged, and sampled; mixing, degassing, diluting, etc. are eliminated. Packers are especially useful for sampling zones of low hydraulic conductivity. Disadvantages of using packers include a possible gain or loss of organic contaminants through sorption or desorption with the packer material and possible unwanted vertical movement of water outside the well, especially when the purging rate is greater than the recovery rate.

In-line parameter monitoring during purging can reduce purging volumes by quickly identifying when purging is complete. Purge water is run through the monitoring chamber (Figure 15) until parameters such as pH, temperature, and conductivity are stabilized. Sample containers are then filled with no interruption in pumping. In-line monitoring combined with low-flow purging

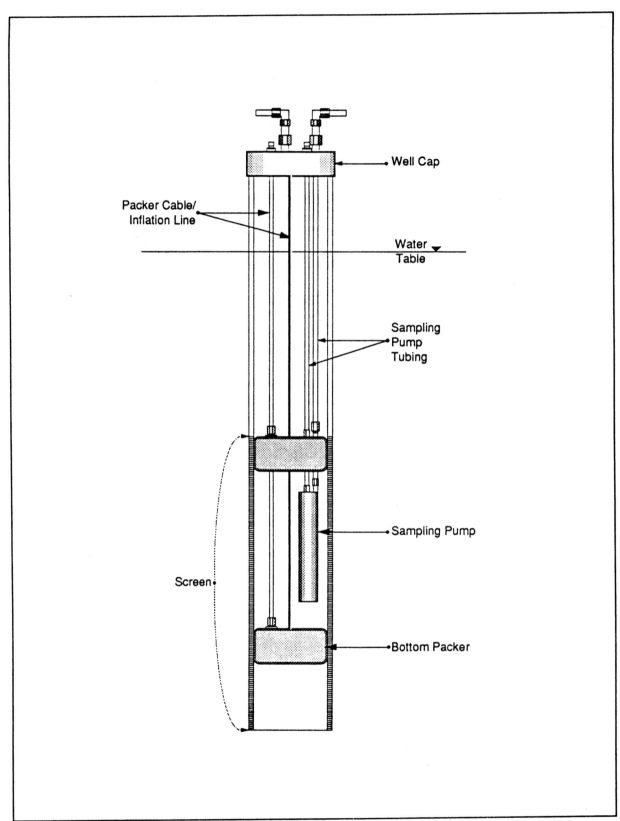

Figure 14 Packer assembly for purge reduction.

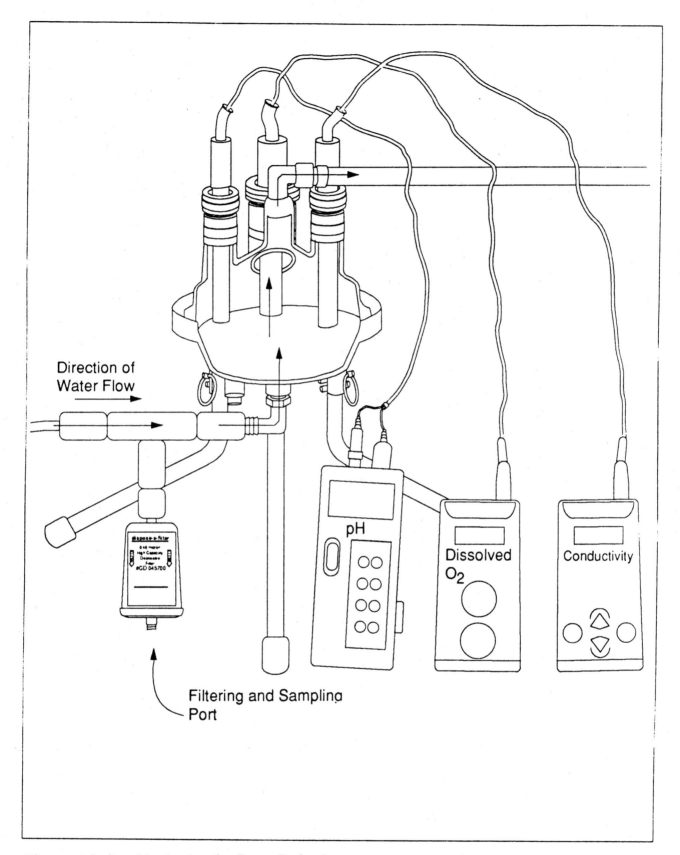

Figure 15 In-line Monitoring for Purge Reduction.

is an effective combination to reduce purging volume.

3.6.3 Determining Casing Volume

Prior to initiating the purging procedure, the volume of standing water in the well must be determined. This water is called a casing volume, bore volume, or well water volume. In most cases, a casing volume is defined as the total volume of standing water in the well casing. However, in some cases, casing volume is defined as the volume of water standing in the well above the top of the well screen. Here it is assumed that the water column within the screened portion of the well is free to interact with the formation water and does not need to be purged.

A manometer or water level meter (Figure 16) is generally used to determine the top and bottom of the water column. This electronic device consists of a water-sensitive stainless steel probe attached to a polyethylene measuring tape usually calibrated in feet and hundredths of feet. The probe is slowly lowered down the well until an audible buzzer (and sometime a light) signals contact with water. The tape is then read at a reference point generally located on the well casing. This is the distance from the reference point to the top of the water column.

The probe is then lowered until there is slack in the tape; this point marks the distance from the reference point to the bottom of the well. By subtracting the two measurements, the height of the water column is determined. However, keep in mind that the reference point on the probe changes during these measurements. The probe reference point for the top of the water column is the hole midway up the probe. The probe reference point for the bottom of the water column is at the bottom of the probe. This distance, generally about 0.15 foot, should be added to the calculated water column length. The water column length is then multiplied by a factor to convert it to gallons of water standing in the well.

Other measuring instruments can be used to determine water column parameters. One instrument is a T-L-C meter (Figure 17) that is designed to measure water column temperature, level, and conductivity. The instrument is set on the conductivity reading and lowered into the well. The conductivity will remain at 0.0 until water is contacted; at that point the reading will jump to some level. This is the top of the water column and the distance is determined by using the calibrated cord. Other parameters are

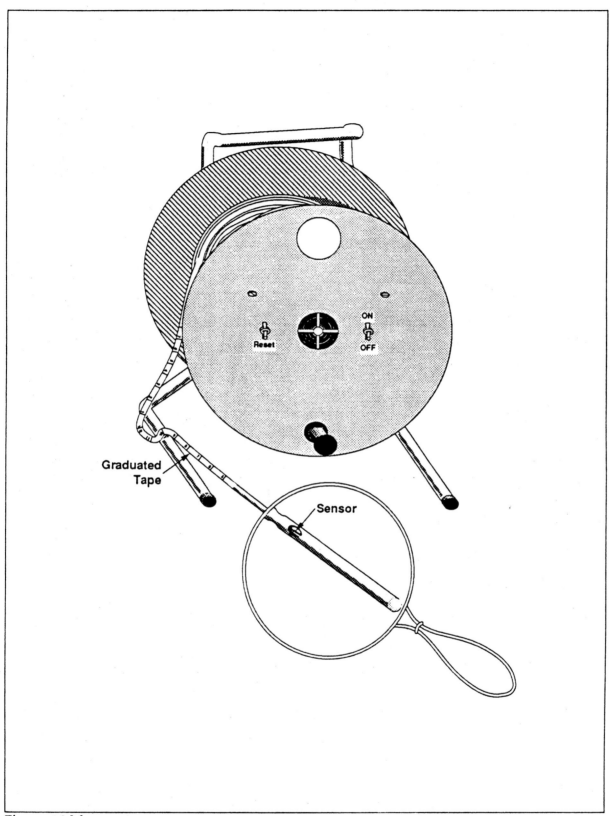

Figure 16 Manometer.

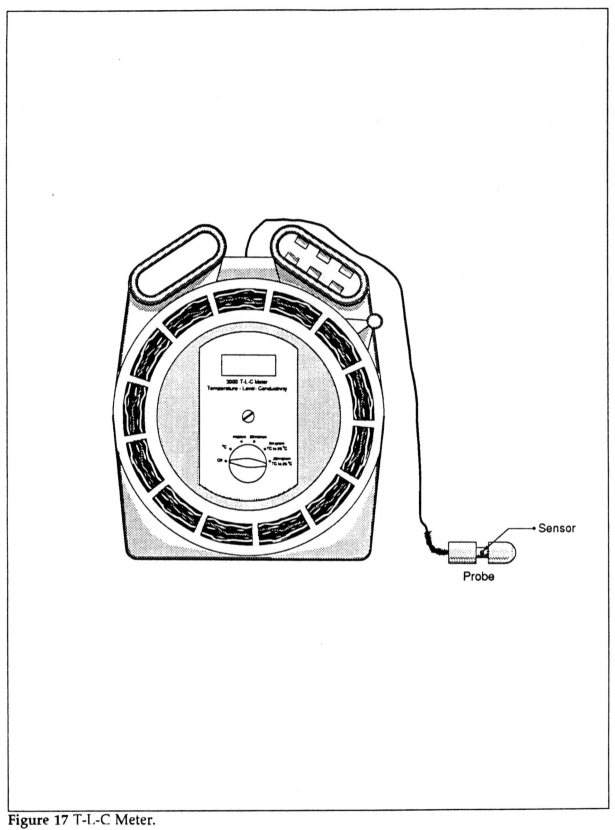

Figure 17 T-L-C Meter.

determined as discussed for the manometer.

Appendix C contains a form commonly used for determining the casing volume, and recording purging and sampling information. At the bottom of the form are factors converting well diameter to gallons per foot.

3.6.4 Sampling Equipment

As with casing materials, all purging, measuring, and sampling equipment inserted into a well must be compatible with the ground water system and sample analysis.

3.6.4.1 Bailers

A bailer is the basic sampling device for ground water investigations (Figure 18). It generally consists of a cylinder with a check ball valve in the bottom and an open top with a cable or rope attachment. When the bailer is lowered into the water column, the force of the water pushes the ball upward and water flows into and out of the bailer. When the bailer is pulled upward, the ball seats in the opening and prevents water from exiting. This column of water is then removed from the well.

Bailers are constructed of stainless steel, Teflon, PVC, polyethylene, and other plastics. Typically, bailers have an outside diameter of 1.66 inches and are 36 inches long; capacity is approximately 1.0 liter. Specialized bailers are also available that have a larger or smaller capacity than the typical bailer. Bailers are available in clear plastic for determining product thickness, and inexpensive disposable bailers are also available. Bailer cable is available in nylon which is used once and disposed of or Teflon coated stainless steel which can be decontaminated and reused.

A variety of top and bottom bailer fittings are used for specialized situations. Most bailers have either an open top or a V-notched top that facilitates pouring water into sample containers. However, the open top design allows for some water exchange when the bailer is pulled up through the water column. If this water exchange is undesirable, a top with a check ball valve that closes during sample retrieval can be used. Attachments are available that allow samples to be discharged from the bottom of the bailer, thus minimizing sample agitation, aeration, and exposure to air.

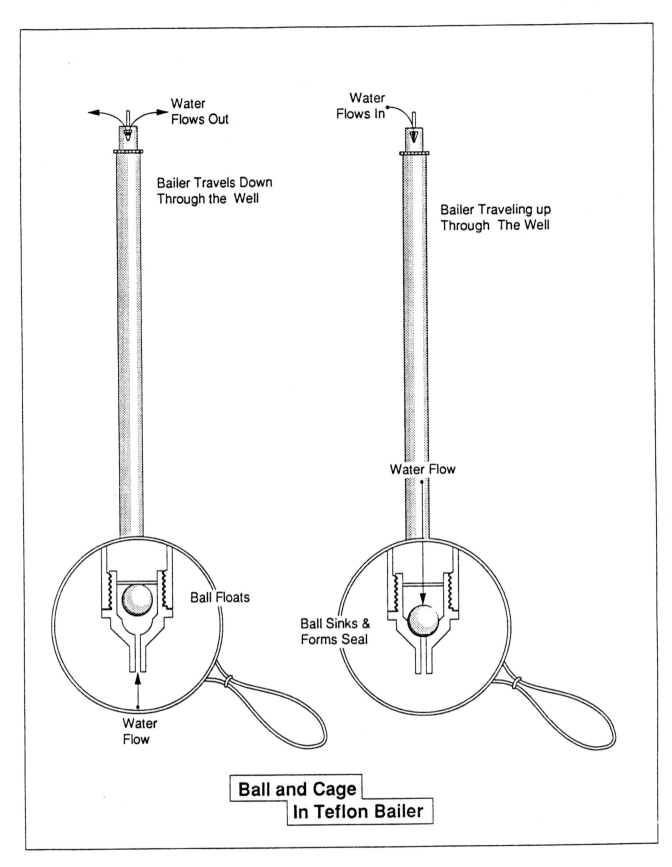

Figure 18 Bailers.

3.6.4.2 Ground Water Pumps

A variety of ground water pumps are available that greatly increase the speed of well purging and sampling. The common ground water monitoring pump types are summarized below. The appropriateness of any particular pump is dependent on several considerations such as analytical parameters, applicable regulations, hydrogeology, and project objectives. The decision on which pump is appropriate for a specific project must be completed prior to sampling.

Suction-Lift Pumps

Suction-lift pumps, or peristaltic pumps, operate by use of a rotating cylinder head that squeezes a flexible tube (Figure 19). The sequential squeezing action lifts water from a maximum depth of 27 feet at sea level. Peristaltic pumps are suitable for removing water from shallow wells. These units are highly portable, relatively low cost, can be used in wells of any diameter, and the water has contact only with the tubing.

These pumps are generally made of heavy anodized aluminum and require an external 12 volt DC or 115 volt AC power source. Most units have stepless variable speed control and operate from 60 to 600 rpm; liquid delivery rate is approximately 1.67 ml per revolution. Pumps have reversible flow features that allow backflushing; this feature is particularly useful when filtering for trace metals.

Because the pump operates by mechanical peristalsis, the water sample has contact only with the tubing. Generally, a medical grade silicon tubing is used and often discarded after one use.

Disadvantages of peristaltic pumps include limited sampling depth, possible degassing and loss of volatiles due to a drop in pressure from the application of the suction, need for a power source, and low pumping rates that can substantially increase purging time.

Centrifugal Pumps

This submersible pump (Figure 20) works through a spinning impeller at the top of the pump that throws water to the outside of the pump shaft. Centrifugal action pushes the water upward through a Teflon tube; additional water moves into the pump and

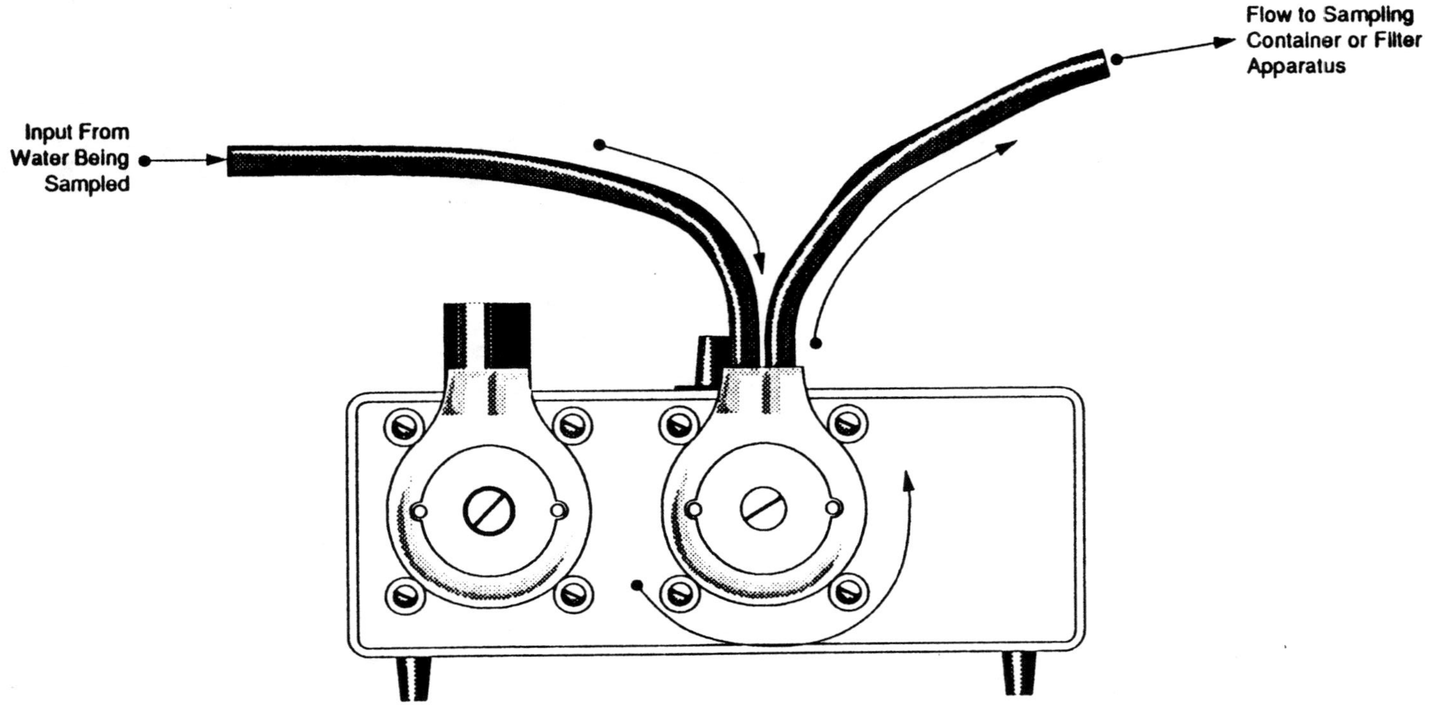

Figure 19 Peristaltic pump.

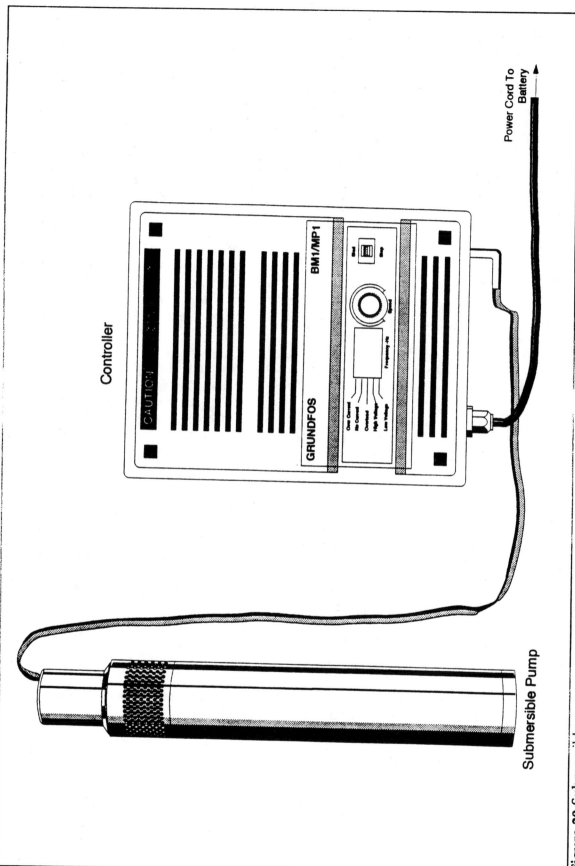

Figure 20 Submersible pump.

the process continues to produce a steady flow of water. The pump is made of stainless steel with Teflon wear and plates and motor leads.

Centrifugal pumps can be operated to a depth of 300 feet (275 foot maximum head). Flow rates range from approximately 100 ml/minute, commonly used for sample collection, to 9 gallons/minute that are often used for purging. Flow rate is controlled by a simple solid state converter. Power is supplied by either a 120 volt or 240 volt generator or battery. The pumps can be as small as 11 inches long and 1.8 inches in diameter; thus, they can readily be used in 2 inch diameter monitoring wells. Pumps weigh between 5 and 6 pounds. Other advantages to centrifugal pumps include ease of use, portability, wide selection of flow rates with little water agitation, and the ability to continuously pump over a long period of time.

Disadvantages associated with centrifugal pumps include a higher initial cost, the need for auxiliary power, possible heating of the water when pumping, frequent gear replacement when sampling wells are high in suspended solids, and possible loss of volatile compounds in the ground water.

Submersible centrifugal pumps are water cooled, but under proper operating conditions, these pumps do not significantly increase water temperatures. During proper operation of these pumps, water moves along the face of the pump to the inlet, thus preventing significant heating of the water. However, when water is pumped vertically into the unit without moving along the pump face (such as using a 1.8 inch diameter pump in a 4 inch diameter well), water temperatures can increase. This heating of the water can be easily prevented by placing a shroud over the pump that forces water to move along the pump face.

Bladder Pumps

Bladder pumps work on the principle of a flexible bladder that is alternately filled and emptied with water (Figure 21). When lowered into a water column, the bottom check valve allows water to flow into a flexible Teflon bladder inside of a stainless steel case. Water fills the bladder, then compressed air is applied through a logic unit that controls the charge and exhaust cycle of the compressed air source. When the compressed gas (usually air or nitrogen) squeezes against the bladder, the bottom check valve closes (as in a bailer) and water is forced upward. When the pressure releases, the top check valve prevents water from flowing back into the bladder, and the bottom check valve opens and water fills the bladder. In this manner water is pumped to the surface.

Bladder pumps are made of stainless steel and Teflon; dimensions are approximately 38 inches in length and 1.7 inches in diameter.

Water is transported to the surface via a Teflon hose. A compressed air source is required that is capable of providing 0.5 pounds of pressure for each foot of well depth. Logic units are either electric (variable from 0.3 to 30 seconds on both the charge and exhaust cycles) or pneumatic (variable from 0 to 60 seconds on both cycles). Maximum pumping depth for the electric unit is 400 feet; pneumatic units can pump to a maximum depth of 300 feet.

The air consumption rate with 100 feet of hose is 125 cubic inches per cycle, with five cycles per minute. For each additional 100 feet of hose, add 59 cubic inches. With respect to pressure needs, for each foot of pump depth 0.5 psi is required. An additional 10 psi is needed to assure a satisfactory flow rate.

Disadvantages of bladder pumps include longer sampling times for deep wells due to the large volumes of gas needed and longer cycle times. Also, check valves can fail when the water contains high amounts of suspended solids, initial costs can be high, and compressed air is needed. In some cases, flow rates are intermittent rather than continuous as is often desired when sampling for volatile compounds.

Dedicated Systems

In some sampling situations, especially when long-term monitoring is to be conducted, a pump is dedicated to a particular well by permanently installing it within the well. Advantages to dedicated systems include elimination of cross-contamination, consistency between sampling events, no re-sampling, and faster set up, purging, and sampling. The primary disadvantage includes high initial cost.

In some ground water monitoring situations, sampling efficiency and integrity require a combination of pumps that are dedicated to a particular well (Figure 22). For example, a centrifugal pump can be used for purging and a bladder pump for sampling. Numerous well configurations are possible, depending on aquifer characteristics and sampling objectives.

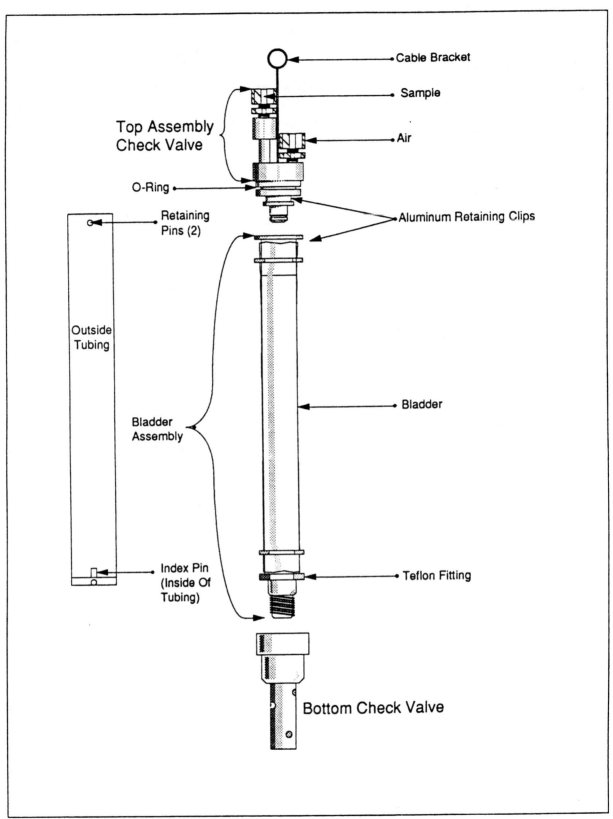

Figure 21 Bladder Pump.

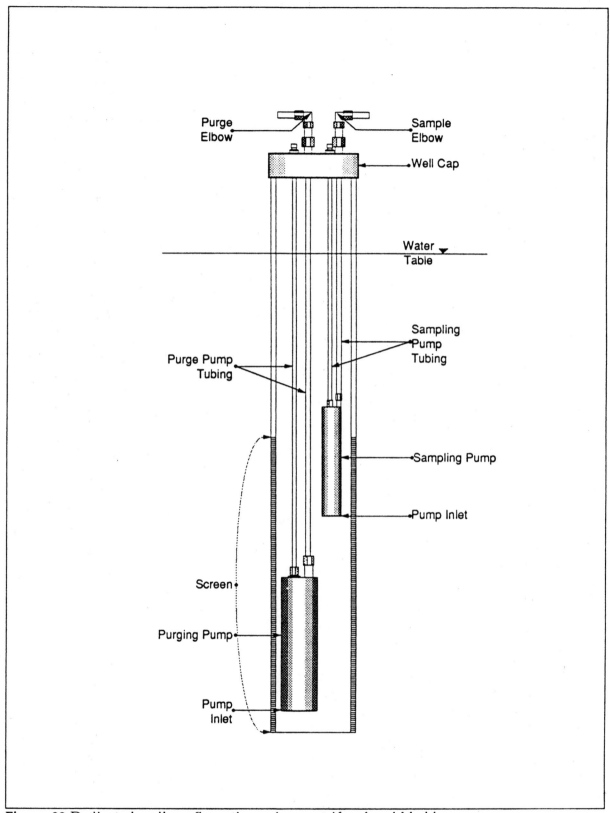

Figure 22 Dedicated well configuration using centrifugal and bladder pumps.

3.6.5 Filtration

Suspended solids in ground water can adsorb ions in the water, thus decreasing the concentration of dissolved constituents in the ground water. The reverse can also happen where particles desorb ions into the water and increase the concentration of dissolved constituents. Therefore, if the analytical program is looking at only dissolved ions in the ground water, removal of particulates is necessary to preserve the sample. Filtering is the common means to remove particulates. Other reasons for filtering ground water samples include determining the percent of suspended solids, separate analysis of suspended solids and filtered water, and removal of solids that may interfere with laboratory analysis or damage laboratory equipment. However, samples collected for organic analysis are not generally filtered.

However, filtering has disadvantages. Filtration commonly involves pushing or pulling a sample through a filter medium by applying either a vacuum or positive pressure to the sample. Undesirable effects of such filtering include one or more of the following:

- change of the partial pressure of dissolved gases in the sample;

- loss of volatile compounds due to exposure to the atmosphere or pressure changes;

- sample oxygenation that can cause precipitation of adsorption of metals that were in solution;

- removal of constituents that are only slightly soluble, such as PCBs and polynuclear aromatic hydrocarbons; and

- sample contamination, such as phosphorus, introduced by the filter material.

Filtering is commonly used when samples are to be analyzed for dissolved metals, when the program focuses on suspended solids and chemicals attenuated by the solids, when monitoring programs are sensitive to changes in sample chemistry due to varying amounts of suspended particles between sampling events, and when programs require precise distinction between the chemical and physical characteristics of dissolved and particulate fractions.

When filtered samples are collected, non-filtered samples are commonly collected at the same time.

Commonly used field filtering systems are described below. Most systems use a 0.45 um glass microfibre or cellulose membrane filter. All of the systems described below do not require electricity, although electric pumps for in-line filtration can be used.

3.6.5.1 Vacuum Filters

With a vacuum filter, the water is pulled through a filter. A common type of field vacuum filter is the "flex" filter (Figure 23). A 0.45 um membrane filter is sealed in the upper one-third of a medical grade PVC bag, thus creating a vertically oriented filter. The bag is placed, with the discharge valve at the bottom, into a non-flexible filter cylinder. The top of the bag is flared over the top of the filter cylinder and held in position with a ring. Water is poured into the filter bag and a hand pump draws a vacuum on the air tube located at the base of the cylinder. This causes the PVC bag to expand against the sides of the cylinder. This expansion creates a pressure differential within the lower portion of the bag and the sample is drawn from the upper portion of the bag, through the filter, and into the lower portion of the bag. Once the filtering is complete, the bag is removed and the sample transferred to a sample bottle by opening the discharge valve at the bottom of the bag.

Flex filters simplify decontamination and decrease the chance of cross-contamination.

3.6.5.2 Pressure Filters

A pressure filter applies pressure to the water sample to push it through the filter. A manual barrel filter is a common type of field pressure filter unit (Figure 24). The water sample is transferred from the bailer (or other collection device) into the barrel reservoir. A filter and filter support screen are set into place, an acrylic plate attached, and the unit is sealed by hand tightening the removable nuts. The unit is then inverted and a sample container placed under the discharge fitting. Pressure is applied with a hand pump to force the sample through the filter and into the collection bottle.

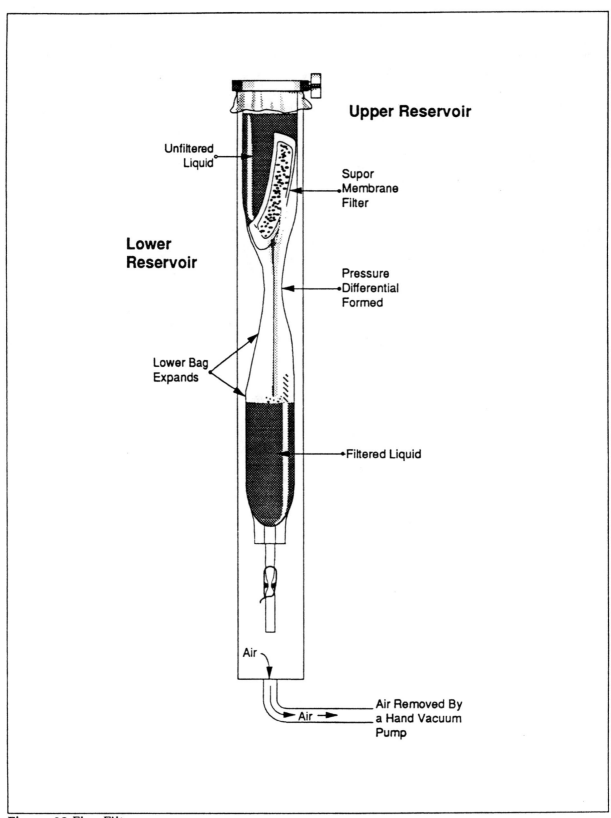

Figure 23 Flex Filter.

3.6.5.3 Disposable In-Line Filter Cartridges

This filtering system consists of a disposable in-line cartridge with inlet and outlet connections that allow for pressure filtration (Figure 24). The filter cartridge is connected directly to the discharge tubing of a peristaltic, bladder, or centrifugal pump. A filtered ground water sample is then quickly and easily obtained.

Advantages to in-line filters include their ability to function at high flow rates, no cross-contamination, reduced sample aeration, less sample exposure to atmospheric conditions, no decontamination, and high throughput in turbid water.

3.6.5.4 In-Line 142mm Backflushing Filter

Another form of in-line filter is the 142mm backflushing filter holder commonly made of acrylic and polycarbonate (Figure 25). It is designed for in-line filtration through a 0.45um filter media. The 142mm diameter surface area allows for rapid filtration of samples through a disposable cellulose acetate or cellulose nitrate membrane. A glass fiber pre-filter may be used to prolong membrane life.

3.7 DECONTAMINATION

All reusable sampling equipment must be decontaminated and non-reusable equipment disposed of in an approved manner. Although site specific procedures must be developed, the following generalized decontamination methods may assist in site specific programs.

When the ground water analytical program focuses on inorganic constituents, decontamination commonly consists of:

- initial soap and water wash;

- dilute acid (10 percent nitric or hydrochloric) rinse; and

- deionized water rinse.

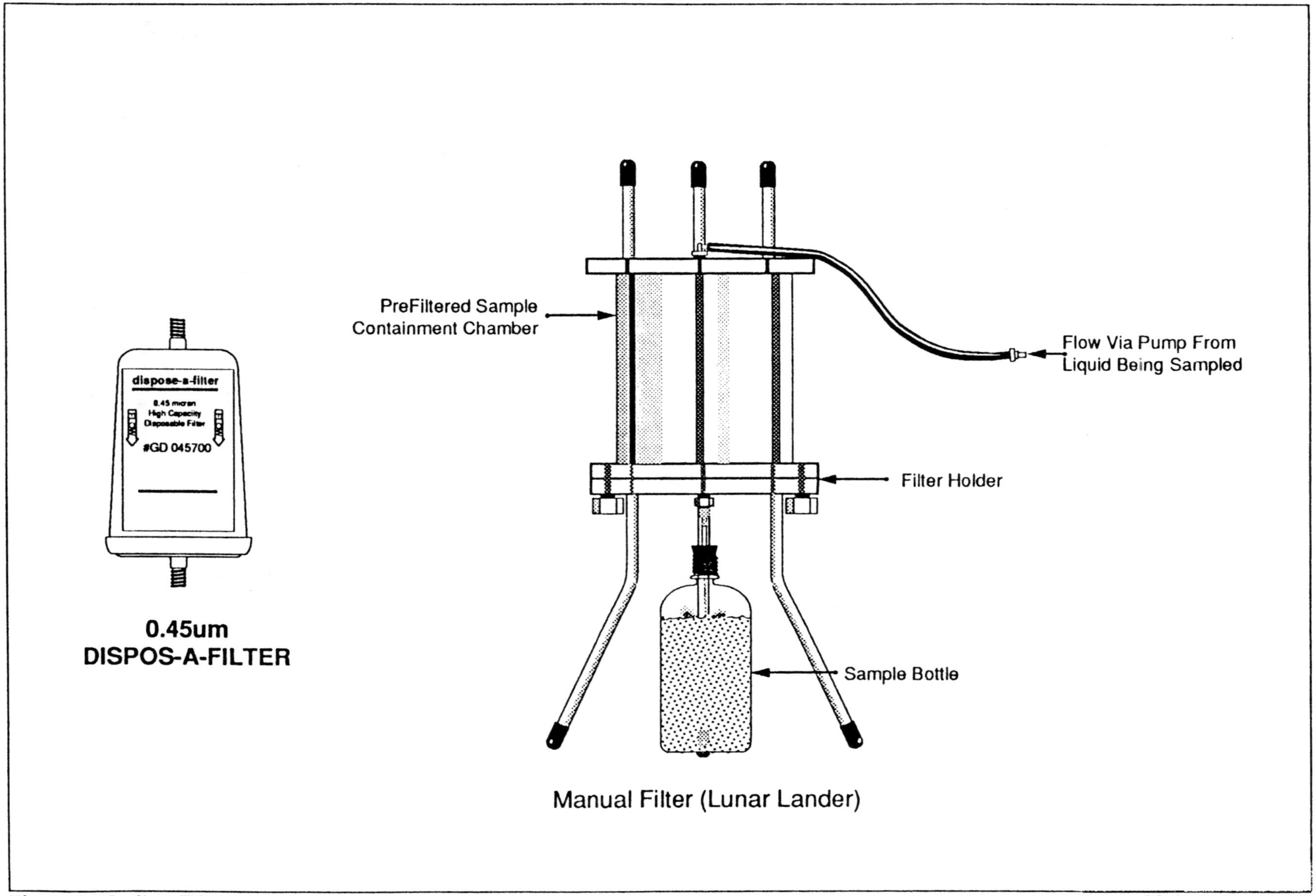

Figure 24 In-line Filters.

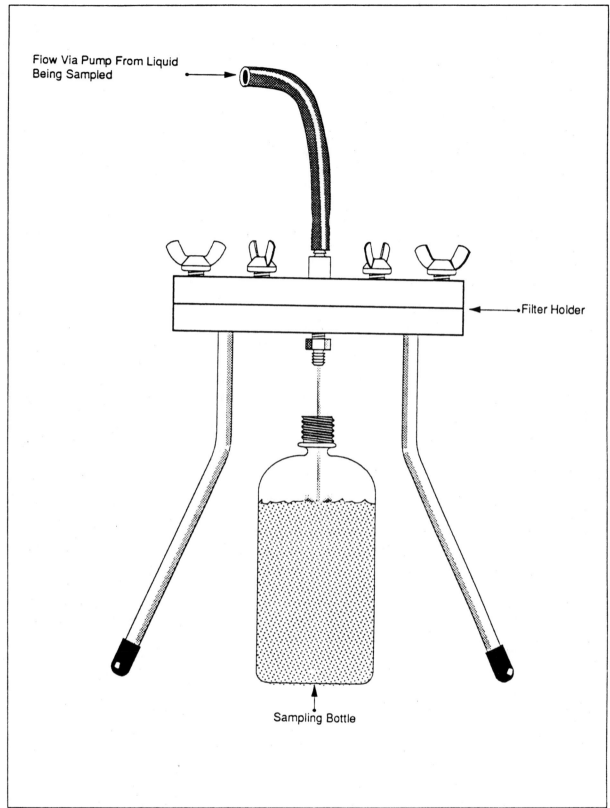

Figure 25 In-line 142mm Blackflushing Filter Holder.

When the ground water analytical program focuses on organic constituents, decontamination commonly consists of:

- initial soap and water rinse;
- solvent rinse (acetone, hexane, isopropyl alcohol, methanol): and
- deionized water rinse.

Both methods often employ pressurized solutions and brushes. Decontamination commonly takes place before and after sampling.

3.8 QUALITY ASSURANCE AND QUALITY CONTROL

A detailed quality assurance and quality control (QA/QC) program should be written for all ground water sampling operations. This document should consider drilling, well development, purging, sampling, and decontamination as previously discussed. Details concerning items such as proper equipment and material selection, appropriate sampling containers, sample preservation and holding times, and applicable SOPs must be fully documented and understood prior to any sampling activities.

Cleanliness during sampling is one of the critical QA/QC components. All down-hole equipment must be thoroughly decontaminated prior to its insertion into the well. During equipment use, precautions must be taken to prevent contamination. This includes wearing gloves, covering the well head and surrounding the area with plastic, placing sampling equipment in clean buckets when not in use, and preventing downhole equipment (including the bailer rope) from touching any potentially contaminated surface. Equipment dedicated to a particular well may also require periodic decontamination, consistent with the site-specific sampling protocol.

3.9 SOURCES OF INFORMATION

Davis, H., J.L. Jehn, and S. Smith. 1991. Monitoring Well Drilling, Soil Sampling, Rock Coring, and Borehole Logging. In (D.M. Nielsen, Editor). Practical Handbook of Ground-Water Monitoring. Lewis Publishers. Chelsea, MI.

Environmental Protection Agency. 1986. RCRA Ground-Water Monitoring Technical Enforcement Guidance Document. U.S. Government Printing Office. Washington, D.C. Oswer-9950.1.

Environmental Protection Agency. 1988. Guidance on Remedial Actions for Contaminated Ground Water at Superfund Sites. EPA Office of Emergency and Remedial Response. Washington, D.C. EPA/540/G-88/003.

Environmental Protection Agency. 1991. Dense Nonaqueous Phase Liquids. Ground Water Issue, March 1991. Office of Solid Waste and Emergency Response. US EPA. Washington, D.C. EPA/540/4-91-002.

Environmental Protection Agency. 1989. RCRA Facility Investigation (RFI) Guidance. Volume II of IV. Waste Management Division, Office of Solid Waste. US EPA. Washington, D.C. EPA 530/SW-89-031.

Heath, R.C. 1989. Basic Ground-Water Hydrology. U.S. Geological Survey Water-Supply Paper 2220. U.S. Geological Survey, Denver, CO.

Herzog, B.L., J.D. Pennino, and G.L. Nielsen. 1991. Ground-Water Sampling. In (D.M. Nielsen, Editor). Practical Handbook of Ground-Water Monitoring. Lewis Publishers. Chelsea, MI.

Herzog, B.L. and W.J. Morse. 1990. Comparison of Slug Test Methodologies for Determination of Hydraulic Conductivity in Fine - Grained Sediments. In (D.M. Nielsen and A.I. Johnson, Editors). Ground Water and Vadose Zone Monitoring. ASTM. Philadelphia, PA.

Kill, D.L. 1990. Monitoring Well Development--Why and How. In (D.M. Nielsen and A.I. Johnson, Editors). Ground Water and Vadose Zone Monitoring. ASTM. Philadelphia, PA.

NOTES

Makeig, K.S. 1991. Regulatory Mandates for Controls on Ground-Water Monitoring. in (D.M. Nielsen, Editor). Practical Handbook of Ground-Water Monitoring. Lewis Publishers. Chelsea, MI.

Nielsen, D.M. and R. Schalla. 1991. Design and Installation of Ground-Water Monitoring Wells. In (D.M. Neilsen, Editor). Practical Handbook of Ground-Water Monitoring. Lewis Publishers. Chelsea, MI.

Nielson, D.M. and A.I. Johnson. 1990. Ground Water and Vadose Zone Monitoring. ASTM. Philadelphia, PA.

Nielsen, G.L. 1991. Decontamination Program Design for Ground-Water Monitoring Investigations. In (D.M. Neilsen, Editor). Practical Handbook of Ground-Water Monitoring. Lewis Publishers. Chelsea, MI.

Sara, M.N. 1991. Ground-Water Monitoring System Design. In (D.M. Nielsen, Editor). Practical Handbook of Ground-Water Monitoring. Lewis Publishers. Chelsea, MI.

Sevee, J.E. and P.M. Maher. 1990. Monitoring Well Rehabilitation Using the Surge Block Technique. In (D.M. Nielsen and A.I. Johnson, Editors). Ground Water and Vadose Zone Monitoring. ASTM. Philadelphia, PA.

Smith, S.A. 1990. Monitoring Well Drilling and Testing in Urban Environments. In (D.M. Nielsen and A.I. Johnson, Editors). Ground Water and Vadose Zone Monitoring. ASTM. Philadelphia, PA.

APPENDIX A

Colorado Well Permit Form

WRJ-5-Rev. 76

COLORADO DIVISION OF WATER RESOURCES
818 Centennial Bldg., 1313 Sherman St., Denver, Colorado 80203

PERMIT APPLICATION FORM

Application must be complete where applicable. Type or print in BLACK INK. No overstrikes or erasures unless initialed.

FOR:
( ) A PERMIT TO USE GROUND WATER
( ) A PERMIT TO CONSTRUCT A WELL
( ) A PERMIT TO INSTALL A PUMP

( ) REPLACEMENT FOR NO. _____
( ) OTHER _____
WATER COURT CASE NO. _____

(1) APPLICANT - mailing address

NAME _____

STREET _____

CITY _____
      (State)   (Zip)

TELEPHONE NO. _____

(2) LOCATION OF PROPOSED WELL

County _____

_____ ¼ of the _____ ¼, Section _____

Twp. _____ ___, Rng. _____ ___, _____ P.M.
       (N,S)   (E,W)

(3) WATER USE AND WELL DATA

Proposed maximum pumping rate (gpm) _____

Average annual amount of ground water to be appropriated (acre-feet): _____

Number of acres to be irrigated: _____

Proposed total depth (feet): _____

Aquifer ground water is to be obtained from:
_____

Owner's well designation _____

GROUND WATER TO BE USED FOR:

( ) HOUSEHOLD USE ONLY - no irrigation (0)
( ) DOMESTIC (1)   ( ) INDUSTRIAL (5)
( ) LIVESTOCK (2)   ( ) IRRIGATION (6)
( ) COMMERCIAL (4)   ( ) MUNICIPAL (8)

( ) OTHER (9) _____
    DETAIL THE USE ON BACK IN (11)

(4) DRILLER

Name _____

Street _____

City _____
      (State)   (Zip)

Telephone No. _____ Lic. No. _____

FOR OFFICE USE ONLY: DO NOT WRITE IN THIS COLUMN

Receipt No. _____ / _____

Basin _____ Dist. _____

CONDITIONS OF APPROVAL

This well shall be used in such a way as to cause no material injury to existing water rights. The issuance of the permit does not assure the applicant that no injury will occur to another vested water right or preclude another owner of a vested water right from seeking relief in a civil court action.

APPLICATION APPROVED

PERMIT NUMBER _____

DATE ISSUED _____

EXPIRATION DATE _____

_____
(STATE ENGINEER)

BY _____

I.D. _____ COUNTY _____

(5) THE LOCATION OF THE PROPOSED WELL and the area on which the water will be used must be indicated on the diagram below. Use the CENTER SECTION (1 section, 640 acres) for the well location.

```
+—-+—-+—-+—-+—-+—-+—-+—-+—-+
|       |←—— 1 MILE, 5280 FEET ——→|       |
+   +   +   +   +   +   +   +   +   +
|       |       NORTH SECTION LINE        |
+—-+— +———+———+———+———+—-+—-+
|       |   |   |   |   |   |           |
↑NORTH  W   |   |   |   |   E           |
+   +   E —+— —+— —+— —+— A   +   +
|       S   |   |   |   |   S           |
|       T   |   |   |   |   T           |
+   +   —+— —+— —+— —+— —+—   +   +
|       |   |   |   |   |               |
+—-+— +———+———+———+———+—-+—-+
|       |       SOUTH SECTION LINE        |
+   +   +   +   +   +   +   +   +   +
|       |       |       |               |
+—-+—-+—-+—-+—-+—-+—-+—-+—-+
```

The scale of the diagram is 2 inches = 1 mile
Each small square represents 40 acres.

WATER EQUIVALENTS TABLE (Rounded Figures)
An acre-foot covers 1 acre of land 1 foot deep
1 cubic foot per second (cfs) . . . 449 gallons per minute (gpm)
A family of 5 will require approximately 1 acre-foot of water per year.
1 acre-foot . . . 43,560 cubic feet . . . 325,900 gallons.
1,000 gpm pumped continuously for one day produces 4.42 acre-feet.

(6) THE WELL MUST BE LOCATED BELOW by distances from section lines.

_____ ft. from _____ sec. line
(north or south)

_____ ft. from _____ sec. line
(east or west)

LOT_____ BLOCK _____ FILING # _____

SUBDIVISION _____

(7) TRACT ON WHICH WELL WILL BE LOCATED Owner: _____

No. of acres _____. Will this be the only well on this tract? _____

(8) PROPOSED CASING PROGRAM
Plain Casing

_____ in. from _____ ft. to _____ ft.

_____ in. from _____ ft. to _____ ft.

Perforated casing

_____ in. from _____ ft. to _____ ft.

_____ in. from _____ ft. to _____ ft.

(9) FOR REPLACEMENT WELLS give distance and direction from old well and plans for plugging it:

_____
_____
_____
_____
_____

(10) LAND ON WHICH GROUND WATER WILL BE USED:

Owner(s): _____ No. of acres: _____

Legal description: _____

(11) DETAILED DESCRIPTION of the use of ground water: Household use and domestic wells must indicate type of disposal system to be used.

_____
_____

(12) OTHER WATER RIGHTS used on this land, including wells. Give Registration and Water Court Case Numbers.

Type or right	Used for (purpose)	Description of land on which used
_____	_____	_____
_____	_____	_____

(13) THE APPLICANT(S) STATE(S) THAT THE INFORMATION SET FORTH HEREON IS TRUE TO THE BEST OF HIS KNOWLEDGE.

_____
SIGNATURE OF APPLICANT(S)

Use additional sheets of paper if more space is required.

Form No.
GAS-35
3/90

STATE OF COLORADO
OFFICE OF THE STATE ENGINEER
818 Centennial Bldg., 1313 Sherman St., Denver, Colorado 80203
(303) 866-3581

INSTRUCTIONS FOR COMPLETING AN APPLICATION FOR A WELL PERMIT

The application must be typed or printed neatly in <u>black ink</u> on an original application form. <u>Copies of application forms cannot be accepted.</u>

At the top of the application, indicate if this is a new, replacement or other type of well. If the well is to be a replacement, supplemental, additional or alternate point of diversion, indicate the permit or registration number of the existing well. If applicable, the Water Court case number should be indicated.

(1) APPLICANT - MAILING ADDRESS
Please complete in full; name, address and telephone number. <u>This is where all correspondence will be sent.</u>

(2) LOCATION OF PROPOSED WELL
Describe the well location only by 1/4 1/4 of a Section, Section, Township, Range and Principal Meridian. This is not necessarily the property description. If you do not know this information, contact your Driller, County Clerk, County Assessor or the County Planning Department to obtain the information from the subdivision maps or survey plat for your land. Be sure the 1/4, 1/4 description corresponds to the location in item no. 6.

(3) WATER USE AND WELL DATA
a) A proposed pumping rate must be indicated. Household, domestic and livestock wells are usually limited to 15 gpm.
b) Generally, proposed <u>average annual appropriation</u> amounts should not exceed the amounts indicated below.

Household use only -	1/3 Acre-Foot	*One acre is approximately 43,560 square feet.*
Domestic (Includes Lawn)-	Variable to 3 acre ft.	*One acre foot is approximately 325,900 gallons.*
Livestock (Cattle) —	1 Acre Foot /100 Head	
Irrigation —	Variable up to 3 Acre Ft./Acre	
Others —	Variable	

c) For irrigation wells. The number of acres you intend to irrigate must be indicated.
d) If this is for lawn and garden irrigation using a domestic well, then indicate one acre, fraction of one acre or square feet. One acre of home garden and lawn irrigation is the maximum allowed. This may be further limited.
e) The proposed well depth must be indicated.
f) Designate the aquifer ground water is to be obtained from. It is required in many cases and especially if more than one aquifer is present. A driller or consultant should be able to assist you with this information.
g) Provide name or number, if any, by which you identify the well.
h) Indicate your proposed use of the well.

Household Use Only -	For use only inside the one single family dwelling (no outside use, lawn or garden irrigation (watering) or stock watering). Under current legislation, some household use wells may include the watering of non-commercial domestic animals.
Domestic —	For no more than 3 single family dwellings (specify the number of homes to be served in item #11), the irrigation of up to one acre of home gardens and lawns (may be further restricted) and the watering of domestic animals.
Livestock —	For watering livestock on farms and ranches.
Commercial —	For use generally in offices or small businesses.
Industrial —	For offices, businesses and manufacturing.
Irrigation —	For the irrigation of cropland.
Municipal —	All uses associated with water districts, subdivision, towns and cities.
Other —	Uses include observation and monitoring wells, and exempt commercial wells.

(4) DRILLER
Proposed driller information is required. Complete in full if you have selected a licensed water well contractor. If you have not selected a contractor, you should indicate 'licensed'. If you plan to construct the well yourself, please indicate this information. Specific rules must be followed if you construct the well yourself. The driller must be licensed at the time of construction.

(5) THE LOCATION OF THE PROPOSED WELL
An 'X' should be placed in the proper 1/4, 1/4 on the map showing the approximate location of the proposed well. For irrigation wells, the land irrigated should be shaded. One regular section contains 640 acres, and is 5,280 feet on a side. A 1/4 section contains 160 acres and is 2,640 feet on a side. A 1/4, 1/4 section contains 40 acres, is 1,320 feet on a side and is part of the information needed in item #2.

(6) THE WELL MUST BE LOCATED BELOW

Distances from the section lines to the proposed well location must be given. <u>Please do not use distances from the lot or property lines.</u> If this information is not known, refer to the instructions in item #2. The distances must correspond to the 1/4, 1/4 description in items 2 and 5.

If your well is to be in a subdivision or a development for 35 acre tracts which are not subdivisions, complete in full. Filing number does not mean recording number. Example: Lot 3, Block 8, Filing No. (Unit) 8.

(7) TRACT ON WHICH WELL WILL BE LOCATED

Complete in full. If the well is to be a domestic or livestock well on a parcel in excess of 35 acres, state the total number of acres being dedicated to the well and provide a legal description if different that item No. 10.

If this is not the only well on the tract (parcel), complete item #12.

(8) PROPOSED CASING PROGRAM

The proposed casing program must be completed. If this information is not known, you should contact a licensed water well contractor or review logs of other wells in the Division of Water Resources records section. A minimum of 4 inch inside diameter casing is required, and at least twenty (20) feet of plain, steel casing must be installed from one (1) foot above the ground surface to 19 feet below the surface.

(9) FOR REPLACEMENT WELLS

If this is an application to replace an existing well, this item must be complete. Plugging and abandoning of the old well is required for all replacement wells. Indicate the distance and direction from the old well to the new location.

(10) LAND ON WHICH GROUND WATER WILL BE USED

Describe the individual tract on which water produced by the well will be used. The complete legal metes and bounds should be included, or provide it as an attachment. If the well is in a subdivision or development, this item should include the lot, block, filing and the name of the subdivision or development. A survey of the property would be helpful. For irrigation wells, describe the total land to be irrigated by this well. Generally, for household, domestic and livestock wells, the acreage stated here should agree with item No. 7.

(11) DETAILED DESCRIPTION

Be specific in describing the use of the water diverted from the well. For household and domestic wells, the type of sewage disposal system must be indicated.

(12) OTHER WATER RIGHTS

Complete this item in full if you have other wells or water rights serving the tract described in (10).

(13) SIGNATURE

The applicant must sign the application. If the applicant is an organization, the signature must be accompanied by the printed or typed name and title of the person signing the application. <u>If the form is signed by someone other than the applicant, it must be accompanied by a letter signed by applicant authorizing that person to sign in the applicant's behalf.</u> Place the date signed next to your signature.

If you have questions, contact the Denver Office or the Division Office of the Division in which your well is located.

DIVISION 1	DIVISION 2	DIVISION 3	DIVISION 4
209 ARIX BUILDING	219 W. 5TH RM. 223	422 4TH ST.	306 S. FIRST ST.
800 8TH AVE.	PUEBLO, CO. 81003	ALAMOSA, CO 81101	MONTROSE, CO. 81402
GREELEY, CO 80631	(719) 542-3368	(719) 589-6683	(303) 249-6622
(303) 352-8712			

DIVISION 5	DIVISION 6	DIVISION 7	DENVER OFFICE
1429 GRAND AVENUE	320 LINCLN AVE STE. E	474 MAIN ST.	RM. 823
GLENWOOD SPGS., CO 81601	STEAMBOAT SPGS., CO. 80477	DURANGO, CO. 81302	1313 SHERMAN ST.(303)
945-5665	(303) 879-0272	(303) 247-1845	DENVER, CO. 80203
			(303) 866-3581

APPENDIX B

Colorado Well Construction and Test Report

FORM NO. GWS-31 11/90

WELL CONSTRUCTION AND TEST REPORT
STATE OF COLORADO, OFFICE OF THE STATE ENGINEER

For Office Use only

1. WELL PERMIT NUMBER _____

2. OWNER NAME(S) _____
Mailing Address _____
City, St. Zip _____
Phone ( ) _____

3. WELL LOCATION AS DRILLED: ____ 1/4 ____ 1/4, Sec. ____ Twp. ____ ____, Range ____ ____
DISTANCES FROM SEC. LINES:
____ ft. from ____ Sec. line. and ____ ft. from ____ Sec. line. OR
(north or south) (east or west)
SUBDIVISION: _____ LOT____ BLOCK_____ FILING(UNIT)____
STREET ADDRESS AT WELL LOCATION:

4. GROUND SURFACE ELEVATION _____ ft.   **DRILLING METHOD** _____
DATE COMPLETED _____. TOTAL DEPTH _____ ft. DEPTH COMPLETED _____ ft.

5. GEOLOGIC LOG:
Depth   Description of Material (Type, Size, Color, Water Location)
_____
_____
_____
_____
_____
_____

6. HOLE DIAM. (in.)   From (ft)   To (ft)
_____   _____   _____
_____   _____   _____
_____   _____   _____

7. PLAIN CASING
OD (in)   Kind   Wall Size   From(ft)   To(ft)
_____   _____   _____   _____   _____
_____   _____   _____   _____   _____
_____   _____   _____   _____   _____

PERF. CASING: Screen Slot Size: _____
_____   _____   _____   _____   _____
_____   _____   _____   _____   _____

8. FILTER PACK:
Material _____
Size _____
Interval _____

9. PACKER PLACEMENT:
Type _____
Depth _____

10. GROUTING RECORD:
Material   Amount   Density   Interval   Placement
_____   _____   _____   _____   _____
_____   _____   _____   _____   _____
_____   _____   _____   _____   _____

REMARKS: _____

11. DISINFECTION: Type _____   Amt. Used _____

12. WELL TEST DATA: ☐ Check box if Test Data is submitted on Supplemental Form.
TESTING METHOD _____.
Static Level _____ ft. Date/Time measured _____, Production Rate _____ gpm.
Pumping level _____ ft. Date/Time measured _____, Test length (hrs.) _____
Remarks _____

13. I have read the statements made herein and know the contents thereof, and that they are true to my knowledge. [Pursuant to Section 24-4-104 (13)(a) C.R.S., the making of false statements herein constitutes perjury in the second degree and is punishable as a class 1 misdemeanor.]
CONTRACTOR _____ Phone (___) _____ Lic. No. _____
Mailing Address _____

Name/Title (Please type or print) _____   Signature _____   Date _____

INSTRUCTIONS FOR WELL CONSTRUCTION AND TEST REPORT

The report must be typed or printed in <u>**BLACK INK**</u>. All changes on the form must be initialed and dated. Attach additional sheets if more space is required. Each additional sheet must be identified at the top by the well owner's name, the permit number, form name/number and a sequential page number. Report depths in feet below ground surface.

This form may be reproduced by photocopy methods, or by computer generation with prior approval by the State Engineer.

The original and one copy of this form must be submitted to the State Engineer's Office within 60 days after completing the well or 7 days after the permit expiration date, whichever is earlier. Another copy of the form must be provided to the well owner.

1. Complete the **Well Permit Number** in full.

2. Fill in **Name and Mailing Address of Well Owner** where correspondence should be sent.

3. Complete the blocks for the **actual** location of the well where drilled. If the owner has more than one well serving this property, provide the identification (**Owner's Designation**) for this well. <u>**DO NOT USE THE OWNER SUPPLIED LOCATION**</u> unless a survey has been provided. For wells located in subdivisions the lot, block and subdivision information must also be provided.

4. Report the ground surface elevation in feet above sea level if available. This value may be obtained from a topographic map. Describe the drilling method used to construct the well and the date completed. Indicate the total depth drilled and the actual completed depth of the well.

5. Fully describe the materials encountered in drilling. Do not use formation names unless they are in conjunction with a description of materials.
   Examples of descriptive terms include:
   Grain size--Boulders, gravel, sand, silt, clay.
   Hardness--Loose, soft, tight, hard, very hard.
   Color--All materials. Most critical in sedimentary rock.
   Depth when water is encountered (if it can be determined).

6. Provide the diameters of the drilled bore hole.

7. The outside diameter, kind, wall thickness and interval of casing lengths must be indicated.

8. Indicate the type and size of filter (gravel) pack and the interval where placed.

9. Indicate the type and setting depth for any packers installed.

10. The density of the grout slurry must be reported and may be indicated as pounds per gallon, gallons of water per sack, total gallons of water and number of sacks used, etc. Specify the grout placement method, i.e. tremie pipe or positive displacement. The percentage of additives mixed with the grout should be reported under remarks.

11. Record the type and the amount of disinfection used, how placed and the length of time left in the hole.

12. Report well test data as required by Rule 10.7. Spaces are provided to report all measurements made during the test. The report should show that the test complied with the provisions of the rules. If a test was not performed explain when it will be done. If available, report clock time when measurements were taken.

13. Fill in **Company Name and Address of Contractor** who constructed the well. The report must be signed by the licensed contractor responsible for the construction of the well.

APPENDIX C

Forms Commonly Used in Well Development and Sampling

TEST BORING REPORT

BORING NO. _____

LOCATION: _____
DRILLER: _____
EQUIPMENT: _____

PAGE NO: _____
LOCATION: _____
ELEVATION: _____
DATE START: _____
DATE FINISH: _____
DRILLER: _____

GROUNDWATER			DEPTH TO:			CASING	SAMPLER	CORE BARREL
DATE	HOURS AFTER COMPLETION	WATER	BOTTOM OF CASING	BOTTOM OF HOLE	TYPE			
					SIZE ID			
					HAMMER WT			
					HAMMER FALL			

DEPTH IN FEET	PID	SAMPLER BLOWS PER 6 INCHES	SAMPLE NUMBER	SAMPLE DEPTH RANGE	FIELD CLASSIFICATION AND REMARKS
−5					
−10					
−15					
−20					
−25					

BLOWS/FT.	DENSITY		BLOWS/FT.	CONSISTENCY	SAMPLE IDENTIFICATION	GROUNDWATER ABBREVIATIONS
0 - 4	VERY LOOSE		0 - 2	VERY SOFT	S - SPLIT SPOON	WD - WHILE DRILLING
5 - 10	LOOSE		3 - 4	SOFT	T - TUBE	NE - NOT ENCOUNTERED
11 - 30	MEDIUM DENSE		5 - 8	MEDIUM STIFF	U - UNDISTURBED PISTON	UR - NOT READ
31 - 50	DENSE		9 - 15	STIFF	G - GRAB SAMPLE	
51+	VERY DENSE		16 - 30	VERY STIFF	X - OTHER	BORING NO.
			30+	HARD	NR - NO RECOVERY	

hazmatt39/boring.rpt

Sheet of

GEOLOGIC BORING LOG

JOB NUMBER: CLIENT: DATE:
BORING NO.: BORING DIAMETER: ELEVATION:
MACH. TYPE: CONTRACTOR:
TEMPERATURE: WEATHER:

DEPTH (FT.)	PID (PPM)	GEOLOGIC DESCRIPTION	SAMPLES NO.	DEPTH (FEET)	SAMPLE TYPE	DRILL PRESS/ BLOW COUNTS	REMARKS

HAZMATT39/GEO1.LOG

WELL DEVELOPMENT LOG

Well Number _____   Job Number _____

Date Started _____   Time Started _____

Date Completed _____   Time Completed _____

Field Personnel _____

Development Method _____

WELL INFORMATION

Description of Measuring point (MP) _____

Total Depth of Well below MP, ft _____

Depth to Water below MP, ft _____

Water Column in Well, ft _____ (Add probe correction)

Gallons in Well ft _____ (gallons/casing volume)

FIELD PARAMETERS

TIME	CASING VOLUME NUMBER	COND. µS/CM	TEMP. °C	pH	COLOR	DRAW-DOWN FT
_____	_____	_____	_____	_____	_____	_____
_____	_____	_____	_____	_____	_____	_____
_____	_____	_____	_____	_____	_____	_____
_____	_____	_____	_____	_____	_____	_____
_____	_____	_____	_____	_____	_____	_____
_____	_____	_____	_____	_____	_____	_____
_____	_____	_____	_____	_____	_____	_____

Comments _____

_____

Note: for a 2 inch diameter well casing, there are 0.16 gallons per foot of water depth.

GROUNDWATER SAMPLING LOG

Project/No_____ Site/Well No._____ Date_____

Site Location_____

Weather_____Time Sampling Began_____Time Sampling Completed_____

EVACUATION DATA

Description of Measuring Point (MP)_____

Height of MP above/Below Land Surface_____MP Elevation_____

Total Sounded Depth of Well Below MP_____Water Level Elevation_____

Depth to Water Below MP_____Diameter of Casing_____

Water Column in Well_____Gallons Pumped/Bailed Prior to Sampling_____

Gallons per Foot _____   Evacuation Method_____

Gallons in Well_____

SAMPLING DATA/FIELD PARAMETERS

Color_____Odor_____Appearance_____Temperature_____ °F/°C

Other (Specific ion; PID, FID etc.)_____

Specific Conductance, umhos/cm_____   pH_____

Sampling Method and Material_____

Constituents Sampled	Container Description	Preservative
_____	_____	_____
_____	_____	_____
_____	_____	_____

Remarks_____

Sampling Personnel_____

WELL CASING VOLUMES

| Gal/Ft 1-1/4" = 0.077 | 2" = 0.16 | 3" = 0.37 | 4" = 0.65 |
| 1-1/2" = 0.10 | 2-1/2" = 0.24 | 3-1/2" = 0.50 | 6" = 1.46 |

Unit 3 Page 166

APPENDIX D

Colorado Well Closure Form

FORM NO. GWS-9 7/89	STATE OF COLORADO OFFICE OF THE STATE ENGINEER 818 Centennial Bldg., 1313 Sherman St., Denver, Colorado 80203 (303) 866-3581	For Office Use only

TYPE OR PRINT IN BLACK INK

WELL ABANDONMENT REPORT

ABANDONED WELL NUMBER IF REGISTERED _____

1. **INDIVIDUAL/COMPANY RESPONSIBLE FOR PLUGGING**

    NAME(S) _____

    Mailing Address _____

    City, St. Zip _____

    Phone (____)_____

2. **ACTUAL WELL LOCATION:** COUNTY _____

    PROPERTY ADDRESS _____
    (Address)   (City)   (State)   (Zip)

    _____ 1/4 _____ 1/4, Sec. _____ Twp. _____ ☐ N. ☐ S., Range _____ ☐ E. ☐ W. _____ P.M.

    Distances from Section Lines _____ Ft. from ☐ N. ☐ S. Line, _____ Ft. from ☐ E. or ☐ W. Line.

3. I (we), report that an existing well was plugged and abandoned for the following reason(s):

    ☐ The well was plugged and abandoned as required under the conditions of approval of Well Permit No. _____.

    ☐ The well was not in use and was abandoned.

    ☐ Other (please explain) _____
    _____

4. The well was plugged and abandoned according to the Water Well Construction and Pump Installation Rules on _____, 19____.

5. The well was plugged with the following materials placed at the indicated intervals:

AMOUNT AND TYPE OF MATERIAL	METHOD OF PLACEMENT	INTERVAL
_____	_____	from _____ feet to _____ feet
_____	_____	from _____ feet to _____ feet
_____	_____	from _____ feet to _____ feet
_____	_____	from _____ feet to _____ feet
INTERVALS OF CASING REMOVED/RIPPED IN FEET		from _____ feet to _____ feet
		from _____ feet to _____ feet

6. I (we) have read the statements made herein and know the contents thereof, and that they are true to my (our) knowledge. [Pursuant to Section 24-4-104 (13)(a) C.R.S., the making of false statements herein constitutes perjury in the second degree and is punishable as a class 1 misdemeanor.]

Name/Title (Please type or print)	Signature	Date

—INFORMATION— It is the responsibility of the well owner to have the well properly plugged and abandoned. The well construction contractor is responsible for notifying the well owner of the plugging and abandonment requirement. This form may be reproduced by photocopy or word processing means.

RULE 11 ABANDONMENT STANDARDS

11.1 General

11.1.1. The plugging and sealing of all wells and test holes is necessary to prevent contamination of ground water and the migration of water through the unused borehole. It is the ultimate responsibility of the well owner to have a well properly plugged and abandoned. The well construction contractor is responsible for notifying the well owner of the plugging and abandonment requirement pursuant to this Rule 11.

11.1.2 Persons authorized to install pumping equipment may plug and abandon wells which do not require the removal of casing from more than one aquifer or the ripping or perforating of casing opposite confining layers.

11.1.3 All materials used for backfilling shall be clean, free from contaminants and chemically inert.

11.2 Unconfined Wells

Wells completed into unconfined aquifers shall be abandoned by filling with either on-site materials, clean sand or gravel to the static water level, then with chemically inert materials to the ground surface. A permanent watertight cover shall be installed at the top of the casing. The casing may be cut off up to five (5) feet below ground level provided the water tight cover is welded or permanently attached to the top of the casing and the hole is backfilled to the land surface.

11.2.1 Cathodic protection holes, dewatering wells, horizontal drains, monitoring and observation holes, percolation holes, piezometer holes, sump pumps and test holes shall be abandoned either pursuant to Rule 11.2 or by removing all casing which was installed and by filling the hole(s) with drill cuttings or chemically inert materials to within five (5) feet of the ground surface. The top five (5) feet of the hole shall be sealed with materials to or less permeable than the top foot of the surrounding soils.

11.3 Confined Wells

11.3.1 Wells which were constructed through more than one aquifer shall be abandoned by placing a grout plug at the confining layer above each aquifer. If records do not show that the casing opposite each confining layer has been grouted when originally installed, the casing shall be either completely removed from the hole, or perforated or ripped opposite such layer prior to placing the grout plug. No plug shall be less than twenty (20) feet in length.

11.3.2 The well casing except for the grout plug intervals shall be completely filled to the land surface with chemically inert materials. A watertight cover will be permanently welded or attached to the top of the casing. The casing may be cut off up to five (5) feet below land surface provided the watertight cover is welded or permanently attached to the top of the casing and the hole is backfilled to the land surface.

— Unit 4 —
Air Sampling

Table of Contents

- 4.1 Regulatory Basis 170
    - 4.1.1 Outdoor Air 170
    - 4.1.2 Indoor Air 170
- 4.2 Sampling Outdoor Air 171
    - 4.2.1 Sampling Locations 172
    - 4.2.2 Sampling Considerations 172
    - 4.2.3 Sample Types 174
    - 4.2.4 Sampling Gases 174
        - 4.2.4.1 Absorption 175
        - 4.2.4.2 Adsorption (Sorption) 175
        - 4.2.4.3 Condensation 178
        - 4.2.4.4 Whole Air Samples 178
        - 4.2.4.5 Direct Analysis 179
    - 4.2.5 Particulates 179
        - 4.2.5.1 Paper Tape Samplers 179
        - 4.2.5.2 Hi-Vol Samplers 182
        - 4.2.5.3 Size Selective Samplers 182
- 4.3 Indoor Sampling 184
    - 4.3.1 Asbestos 185
        - 4.3.1.1 Air Sampling Prior to Abatement 186
        - 4.3.1.2 Air Sampling During Abatement 186
        - 4.3.1.3 Air Sampling Following Abatement 187
    - 4.3.2 Radon 190
    - 4.3.3 Formaldehyde 190
    - 4.3.4 Biogenic Particles 192
- 4.4 Sources of Information 193

— Unit 4 —

AIR SAMPLING

4.1 REGULATORY BASIS

4.1.1 Outdoor Air

The Clean Air Act of 1970 and its numerous amendments form the complex federal regulatory basis for air quality. This act provides for both prevention and control of discharges into the air from stationary sources (such as factories) and mobile sources (such as automobiles).

The two basic strategies used to manage air quality under the Clean Air Act are as follows:

> National Ambient Air Quality Standards--These are standards that specify a level at which air pollutants can be safely tolerated without harm to public health or natural resources. Both primary and secondary air quality standards have been established. Primary standards reflect the allowable concentrations required to protect public health. Secondary standards are more stringent and are set to protect both public health and the environment. Few secondary standards have been set.

> Emission Standards--Pollutants entering the air are regulated through a system of permits for individual discharges, with the amount of pollutants that can be legally discharged depending on the type and age of the source and the health effects of the substance being released. The standards include visible emission standards, prohibitive standards, limitations on fuel content, and numerical standards including the National Emission Standards for Hazardous Pollutants (NESHAP).

4.1.2 Indoor Air

With respect to indoor air quality, federal programs are piecemeal and confusing. Under the Clean Air Act the EPA has authority to regulate indoor air quality. Yet with exception of the regulation of asbestos under the Toxic Substances Control Act (TSCA), the EPA has promulgated few regulations for indoor air quality. OSHA sets and enforces indoor air quality standards for most commercial establishments. The Department of Energy (DOE) is involved in indoor air quality because its energy conservation recommendations might exacerbate indoor air quality problems. Additionally, the use of uranium mill tailings within the foundations of buildings has caused elevated levels of indoor radon. The Consumer Product Safety Council (CPSC) has statutory authority to regulate the use of consumer products that may pose unreasonable health and safety risks to consumers. Under this authority, the CPSC banned the use of asbestos-containing spackling compounds and attempted to ban urea-formaldehyde foam insulation. The Department of Housing and Urban Development has promulgated product standards for plywood used in mobile home construction and has limited the availability of FHA-financed loans for home construction in high-radon areas.

With the lack of an effective regulatory program, nonregulatory strategies have been used for indoor air quality. A variety of federal agencies and professional groups have issued health guidelines that are not enforceable, but carry the force of regulation since the government or scientists believe that levels above the guidelines are unsafe. Such guidelines are used for ventilation, radon, carbon dioxide, chlordane, heptachlor, chlorpyrifos, formaldehyde, and ozone.

4.2 SAMPLING OUTDOOR AIR

Both gaseous and particulate matter must be addressed in an air sampling program. Gaseous pollutants include carbon oxides (such as carbon monoxide and carbon dioxide), sulfur compounds (such as sulfur oxides, sulfur dioxide, and hydrogen sulfide), nitrogen compounds (such as nitrous oxide, nitric oxide, nitrogen dioxide, and ammonia), hydrocarbons and their derivatives, photochemical oxidants (such as ozone), and halogenated hydrocarbons (such as CFCs).

Particulate matter is a general term that includes very small solid and/or liquid particles that are produced by natural sources (such as pollen, smoke, soil particles) and by human activity (soot, fly ash, metal particles). The smaller particles, those less than 1 micron in size, behave as gases while the larger particles act as solids and are strongly affected by gravity.

4.2.1 Sampling Locations

The air monitoring network should provide adequate coverage to characterize both upwind (background) and downwind sites. Four air monitoring zones are generally necessary for initial monitoring. These zones are illustrated in Figure 1 and described as follows:

- upwind (based on the expected prevailing wind flow) of the source and near the site boundary to characterize background air quality;

- downwind at the source boundary;

- downwind at the site boundary; and

- downwind at offsite receptor locations.

Air sampling stations should be located at least 100 meters from any building or other structure that would significantly alter the air flow. Inlet exposure height of the air monitors should be 2 to 15 meters from the ground to represent the potential inhalation exposure but not be unduly biased by road dust and natural wind erosion phenomena.

Subsequent monitoring may be needed to redefine sampling locations based on the initial data collected and the application of those data to dispersion. Dispersion is dependent on wind, turbulence, atmospheric stability, topography, and long-range transport via jet streams and high pressure systems. These variables are generally integrated into complex dispersion models that may define the need for additional sampling and monitoring locations.

4.2.2 Sampling Considerations

Sample size must be sufficient for subsequent analysis; sample size is dependent on the sampling rate and duration. The sampling rate is determined by the collection efficiency of the method used and equipment limitations. For gases, collection efficiency decreases with increasing sampling rates and sampling time is often relatively long to collect sufficient quantities of pollutants for analysis.

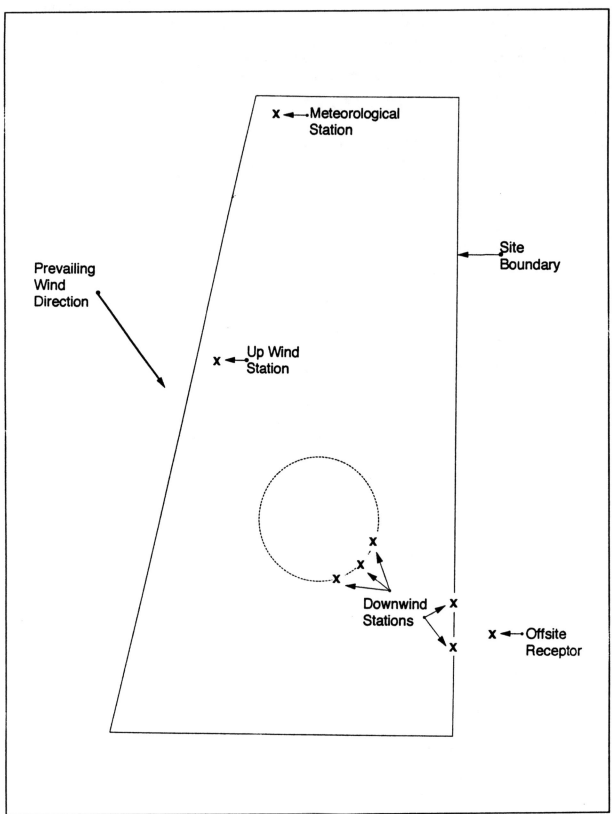

Figure 1 Simple air monitoring network.

Most air quality data requirements specify that samples be averaged over a 24 hour period, especially if long-term evaluations are to be made. In contrast, some data requirements specify real-time data, especially if sampling is for personnel health and safety. Unfortunately, real-time data are so voluminous they are difficult to interpret. Therefore, real-time data are often summarized into hourly averages that can be more readily interpreted.

Meteorological data is necessary for outdoor air monitoring programs. Wind speed, wind direction, relative humidity, and precipitation are the most common parameters monitored. These data are necessary to correctly interpret the air quality measurements.

4.2.3 Sample Types

Sample types include grab and integrated samples. Grab sampling is primarily used for personal monitoring. Here a small volume of air is collected and often analyzed within seconds. Grab sampling can be used when whole air samples are to be collected for later analysis.

Integrated sampling utilizes either static or continuous techniques. Static (or passive) sampling involves the collection of pollutants by diffusion of gases into a collection medium, the sedimentation of large particles into a container, or the impaction of particles on sticky paper or sensitive film. Static methods are simple and low cost, but are of limited use because the collection period is usually 7 to 30 days. However, they are used for personal monitoring and for measurement of some indoor air quality parameters such as radon.

Continuous sampling systems collect a continuous sample of air over a specified period of time, often a 24-hour period; such instruments now dominate air quality sampling and monitoring programs. These sampling systems utilize relatively large collection systems that are often located at stationary points.

4.2.4 Sampling Gases

Gases are collected by absorption, adsorption, or condensation; absorption is the most common system. The majority of hazardous constituents in the air are classified as gaseous or vapor-phase organics.

4.2.4.1 Absorption

In gas absorption, pollutants are drawn into a dispersion tube that is below the surface of an absorption reagent, often an organic solvent (Figure 2). As the gas bubbles through the liquid, the pollutant is dissolved and absorbed into the liquid which is then analyzed. This method is commonly used for aldehydes, ketones, and formaldehyde. Disadvantages include the evaporation of the solvent if large volumes of air are bubbled through it, and that certain compounds are unstable or reactive and will decompose during collection if not changed to more stable forms.

4.2.4.2 Adsorption (Sorption)

In gas adsorption, airborne pollutants are physically attracted to a solid (termed a sorbent) that has a large surface area, such as activated carbon, synthetic polymeric resins such as Tenax, molecular sieves, or silica gel (Figure 3). The adsorbent is then purged from the system and analyzed at the laboratory.

The advantages of sorbent techniques are their ease of use and ability to sample large volumes of air. Commercially available sorbent cartridges can be applied to many applications and easily adapted for portable monitoring pumps. Sorbents can be used singly or in series for maximum retention of airborne pollutants. Finally, sorbents are especially applicable to integrated or long-term sampling because large volumes of air can be passed through the sampling cartridges before breakthrough occurs.

When choosing a sorbent, the following items should be kept in mind.

- Sorbents can be easily contaminated during manufacturing, shipping, or storage; therefore, cleaning with solvents and thermal conditioning are often required prior to use.

- No single adsorbent exists that will retain all gases. The efficiency of retention of a pollutant on a sorbent depends on the chemical properties of both the pollutant and sorbent. Generally, a sorbent that works well for nonpolar organics such as benzene will perform poorly with polar organics such as methanol.

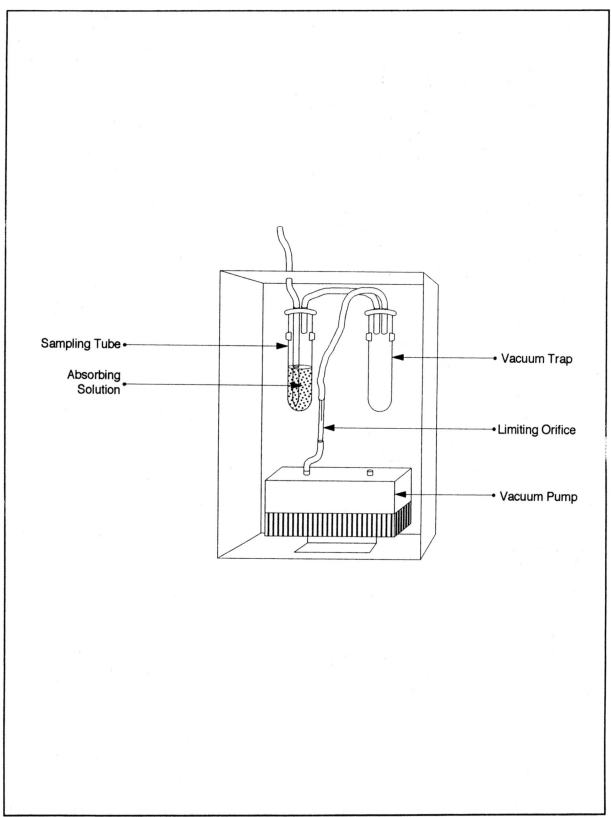

Figure 2 Gas absorption unit.

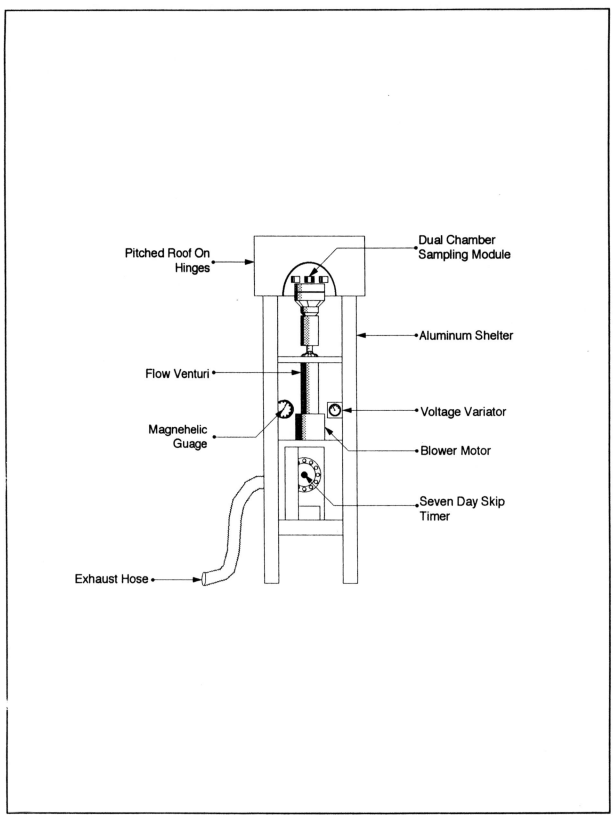

Figure 3 Gas adsorption unit.

- Carbon sorbents include activated carbon, carbon molecular sieves, and carbonaceous polymeric resins. The major advantage of carbon sorbents is their strong affinity for volatile organics, although some organic compounds may become irreversibly adsorbed on the carbon. Carbon adsorbents will retain some water and their use is often restricted in humid conditions.

- Polyurethane foam (PUF) can effectively collect a variety of semivolatile organics, including PCBs and pesticides. This material has the ability to perform at high flow rates, typically in excess of 500 liters per minute, thus minimizing sampling time. In most cases, a filter is placed ahead of the PUF cartridge; this filter traps particles that could plug the PUF cartridge.

In some situations, sorbent trains are used that consist of numerous sorbent traps and backup traps to minimize the potential for breakthrough. The Volatile Organic Sampling Train (VOST) is an accurate method to collect a broad range of organic compounds. It uses dual sorbents and duel in-series traps.

4.2.4.3 Condensation

In gas condensation, air is pumped through a collection container maintained at a sub-ambient temperature. The low temperature of the collection container causes many organic vapors to condense and thus be captured by the container. The liquid is then removed for analysis.

A variation of this system captures volatile organics in stainless steel U-tubes immersed in liquid oxygen or liquid argon. The major advantage of this cryogenic method is that all vapor phase organics, except for the most volatile, are captured. Disadvantages of the cryogenic system include plugging the system with ice in high humidity conditions, the necessity of analyzing the sample all at once thus making multiple analyses of the sample impossible, and the necessity of handling and transporting cryogenic liquids.

4.2.4.4 Whole Air Samples

Air samples can be collected without preconcentrating as is done in the above sampling methods. Samples can be collected in glass or stainless steel containers, or in inert flexible containers such as Tedler bags (Figure 4).

Rigid containers are generally used for collection of grab samples while flexible or rigid containers are used to collect integrated samples. Use of a flexible container to collect whole air samples requires the use of a sampling pump with flow rate controls. The pump hose is connected to the stainless steel sampling port on the bag and the bag filled with air. The sampling port is then closed and the contained air analyzed in the field or in the laboratory. Sampling with a rigid containers is often accomplished with a vacuum bottle. The air from a stainless steel container is evacuated in the laboratory and the valve closed. In the field the valve is opened to the proper flow rate and a grab sample collected over a specified period of time, usually a short interval ranging from a few minutes to an hour. Following sample collection, the valve is closed and the air sample sent to a laboratory for analysis.

4.2.4.5 Direct Analysis

A variety of instruments are available for direct analysis of air samples in the field. Such instruments are commonly used for personal safety, site characterization, and field analysis.

The direct analysis methods does not require preconcentration or separation of air components, and avoids component degradation during storage. Air is drawn directly into the instrument and analyzed.

4.2.5 Sampling Particulates

Although a variety of techniques can be used to collect particulate matter, gravitational settling, filtration, and impaction are the most widely used. The following sampling systems use one or more of these methods to collect particulates.

4.2.5.1 Paper Tape Samplers

This type of sampler pulls ambient air through a cellulose type filter where particles are trapped (Figure 5). After a 2 hour sampling interval, the tape automatically advances and the sampling cycle begins on a clean piece of tape. Many units are equipped with densitometers to determine optical density that is related to light transmission and atmospheric haze. Routine use

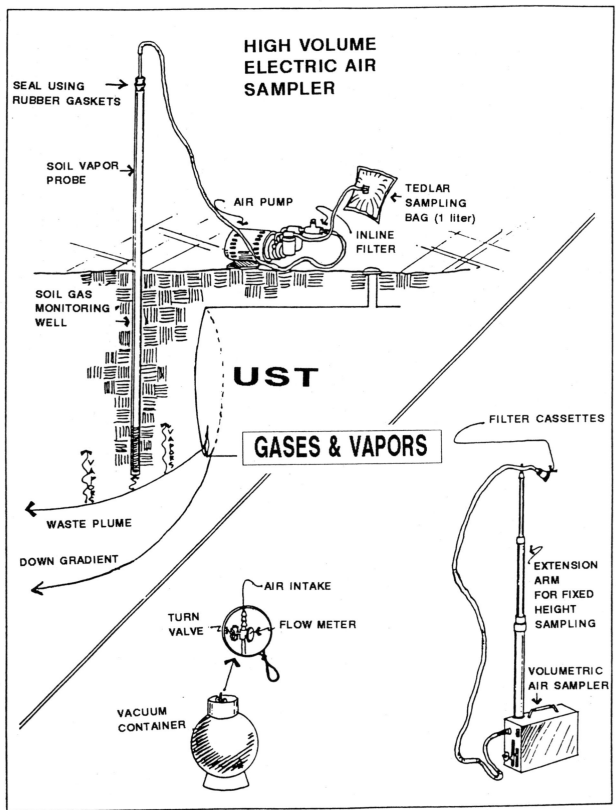

Figure 4 Whole air samplers.

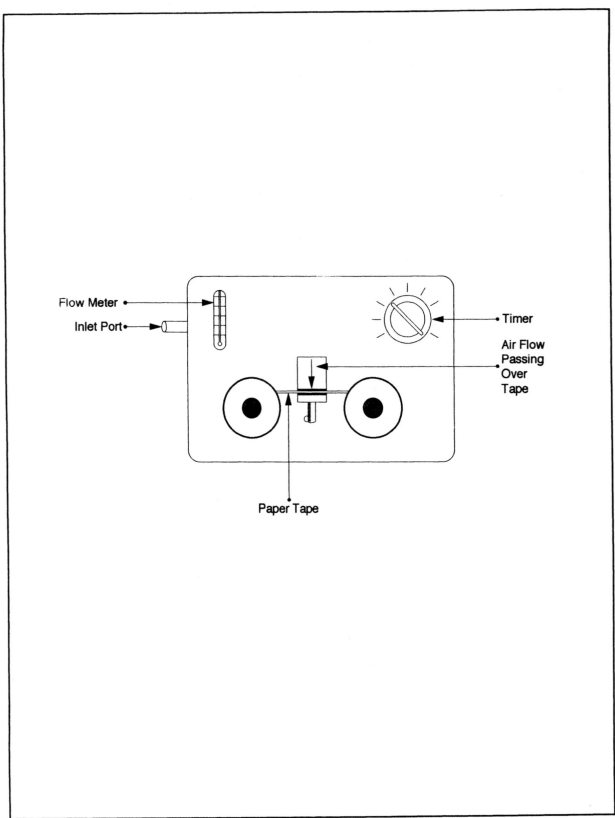
Figure 5 Paper tape particulate sampler.

of paper tape samplers has been discontinued in most locations, although these units are used on a standby basis in many metropolitan areas.

4.2.5.2 Hi-Vol Samplers

A heavy-duty vacuum cleaner type motor pulls an air sample through a glass fiber filter (usually 8 by 10 inches) mounted parallel to the ground. These instruments can collect a 2000 cubic meter sample during a 24-hour period. After the sampling period, the filter is removed and the particles analyzed for physical and/or chemical properties. The sampling system is housed in a shelter that protects the system from wind and precipitation. Hi-vol systems can efficiently collect particles in the size range of 0.3 to 100 microns. Such samplers are normally programmed to collect a continuous 24-hour sample every 6 days. These are the most commonly used particle samplers.

Although glass fiber filters are commonly used for hi-vol systems, organic and membrane filters such as cellulose ester and Teflon can also be used. These filters have greater uniformity in pore size and often lower contamination levels than are found in glass fiber filters.

4.2.5.3 Size-Selective Samplers

This sampling system segregates collected particles into discrete size ranges based on their aerodynamic diameters through the physical principle called inertial impaction. Air is drawn through the sampling unit and deflected from its original flow path (Figure 6). The inertia of the suspended particles causes them to impact a deflecting surface where they are collected on filters according to size. Teflon filters are generally recommended because they exhibit negligible particle penetration and result in a low pressure drop during the sampling period.

Such impactors are now commonly used to meet the recent EPA PM10 standard that requires sampling and separation of particulates below 10 microns in size. Both cascade and dichotomous impactors are used to meet the performance specifications.

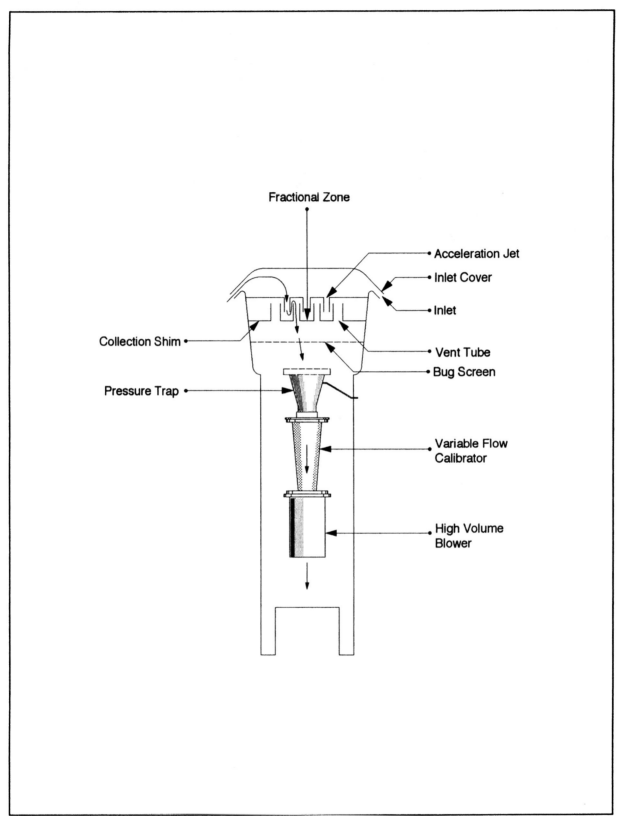

Figure 6 Size-selective sampler.

4.3 INDOOR SAMPLING

Studies of human activity found that approximately 22 hours per day are spent indoors and, therefore, humans receive most of their personal pollution exposure indoors. Such exposures are probably of more public health importance than exposures to polluted ambient air. Some sources of indoor air pollution in commercial buildings are as follows:

- trinitroflurenone from laser printers;

- ammonia and acetic acid from blueprint machines;

- ozone from copying machines;

- butyl methacyleate from signature machines;

- formaldehyde from carbonless copy paper;

- pesticides from insect control;

- combustion gases from cafeterias and laboratories;

- boiler additives such as diethyl ethanolamine;

- organic vapors from rugs and furniture;

- cigarette smoke;

- microbial contamination from ventilation systems; and

- radon from subfoundation materials

These pollutants often combine to create "sick building syndrome." This is where a high number of individuals complain of nonspecific ailments including headaches, unusual fatigue, eye, nose, and throat irritation, shortness of breath, and nausea.

Specific symptoms sometimes occur, such as Legionnaires' Disease. Air quality monitoring to pinpoint the cause of the problem has often proven to be fruitless because contaminant levels are usually considered to be too low to have caused the illness. Physiological factors may also contribute to building-related health problems.

When sampling for indoor air contaminants, it is not appropriate to test under conditions that produce results reflecting minimum concentrations. Sampling with doors and/or windows open will produce meaningless data. Accurate assessment of health effects from indoor pollutants requires sampling under worst case conditions. Sampling usually requires building closure to approximate worst case conditions and to reduce variability associated with occupant practices.

Following is a discussion of the more common indoor air problems for which sampling techniques have been recommended.

4.3.1 Asbestos

Asbestos is a collective term used for a variety of asbestiform materials that include chrysotite, anthphylite, riebeckite, cummingtonite-grunerite, and actinolite-tremolite. However, chrysotite accounts for approximately 95 percent of the asbestos sold. Asbestos has been used in over 3000 commercial products.

Documented health hazards from asbestos exposure include lung cancer, mesothelioma, asbestosis, and nonmalignant pleural disease. Asbestos poses a significant health risk even at low exposure levels that can be found within many buildings. Children are especially at risk because exposure at a young age greatly increases the lifetime risk of mesothelioma.

Although asbestos is considered a ubiquitous indoor contaminant, certain forms and certain uses of asbestos greatly increase the risk of exposure. Friable asbestos (asbestos that can be crushed by the application of hand pressure) poses the greater health risk than non-friable asbestos. Friable asbestos products that have been sprayed or troweled on surfaces, often ceilings and beams, probably pose the most serious health hazard. Fiber release from such materials occurs as they age and as the building materials weaken. This can result in continuous fiber erosion or a delamination from the substrate. Asbestos has been commonly used as thermal insulation around steam lines and boilers. This material poses less of a risk because it is usually covered with a rigid cloth or paper material and it is often inaccessible to most building occupants. Floor coverings may also contain asbestos fibers, but rarely pose a significant risk unless occupant traffic is unusually high.

Asbestos samples are collected by drawing a known volume of air through a filter. The filter is housed in a plastic cassette that is attached to the sampling pump with flexible tubing. Pumps are either electric (plug in) or battery powered and calibrated to draw a known volume of air through the filter over a given period of time. Filters are commonly made of a mixed cellulose ester (MCE).

Sampling procedures for asbestos fibers are specified under the NESHAP regulations. Sampling is required before, during, and after asbestos remediation. However, sampling regulations frequently change, so be sure all sampling procedures follow the most recent regulations.

4.3.1.1 Air Sampling Prior to Abatement

Air sampling is conducted before abatement activities to estimate the existing airborne fiber concentrations inside and outside the building; this sampling is termed prevalent level sampling. Such sampling is particularly useful when an abatement project is conducted in a portion of the building with other areas of the building remaining occupied. Airborne fiber levels monitored in these occupied areas during the abatement project should never exceed the indicated prevalent level in these areas before the project began.

Low asbestos concentrations are typically found prior to abatement activities, therefore, a large volume of air should be sampled to obtain a representative sample. Typically, 3000 to 4000 liters of air are sampled using pumps that have a flow rate of 2 to 10 liters per minute.

Prevalent level samples should be collected throughout the building as well as in the areas where abatement will take place. Generally, one sample should be taken for every 5000 square feet of floor space. In addition, at least two samples should be collected from outside the building.

4.3.1.2 Air Sampling During Abatement

Personal air sampling is conducted during remediation by using a low volume pump (flow rate of 0.5 to 2 liters per minute) that is attached to the worker's belt and using the previously mentioned filter. At least 25 percent of the work force should be monitored on a periodic basis during remediation. Samples are generally collected over an eight-hour period; filters may need to be changed several times during this period to prevent overloading.

In addition to personal samples, area air samples are collected inside the work area on a daily basis. Two to three samples of 60 to 120 liters of air are usually adequate to index the airborne fiber concentrations inside the work area. Air samples are usually collected in different places within the abatement area from one day to the next. Typical sample locations are in the immediate vicinity of removal, upwind of the abatement area, downwind of the abatement air, in the dirty equipment room and near critical barriers. Small personal sampling pumps can be used for sample collection because the sampling of large volumes of air is not necessary.

During abatement, air samples are collected from locations outside the work area, but inside the building to determine the effectiveness of fiber containment. Sampling is also conducted outside the building to detect leakage of fibers from the worksite. In most cases, the collection of 3000 to 4000 liters of air is necessary for a representative sample.

4.3.1.3 Air Sampling Following Abatement

Area air sampling is conducted upon conclusion of the abatement project to determine the airborne concentration of residual fibers. Samples are collected using aggressive techniques. This involves physically or mechanically agitating the air in the work area during sampling. Typically, a one horsepower leaf blower is used on all surfaces in the work area to dislodge any residual fibers. Then a standard box fan is left on for the duration of the sampling period to keep any dislodged fibers airborne. In most cases, high volume pumps are used to collect the samples.

4.3.2 Radon

As an indoor air contaminant, radon is unique because it is a naturally occurring contaminant; humans simply trap it and allow the gas to accumulate to dangerous level. Radon is produced from the radioactive decay of radium found in uranium ores, phosphate rock, granite, and other common minerals and is usually transported as a gas in soil.

Radon is inert and harmless; however, it undergoes radioactive decay and produces a number of short-lived progeny that emit alpha particles. The progeny are electrically charged and can attach to airborne particles that are subsequently inhaled. Once in the respiratory airways, the progeny become lodged and continue to emit alpha particles as the decay process continues. Cancer can result from long-term exposure to elevated levels of radon.

Over 90 percent on indoor radon is from soil gas that is transported into buildings by pressure-induced convective flows through foundation cracks and other openings. Water and masonry materials are other sources of radon in buildings.

Initial screening for radon can be accomplished by a number of inexpensive sampling methods as follows:

Charcoal Canisters

This sampling method uses charcoal contained in a metal or plastic container. Following exposure for two to seven days, the canister is removed and placed in a gamma ray detector. Gamma ray energies associated with the decay of radon progeny are detected and quantified.

Electret Monitors

This method employs a disk of Teflon that has been given a stable electrostatic charge. The electret disk is suspended in a plastic chamber (Figure 7). Air, and any radon, enters the chamber through a particulate trap. The subsequent decay of radon produces charged progeny. The negatively charged progeny are attracted to the positively charged electret and cause a voltage drop. Radon concentration is determined by the change in electrical charge on the electret using a specifically designed voltmeter. This method can be used for both short-term testing (less than 90 days) and long-term testing (more than 90 days).

Track-Etch Detectors

A piece of sensitized film is placed in a canister and air allowed to circulate over the film. As the alpha particles associated with radon progeny pass over the film, they etch the film. These etchings are then counted microscopically and converted to radon concentrations. This method can be used for both short and long-term testing.

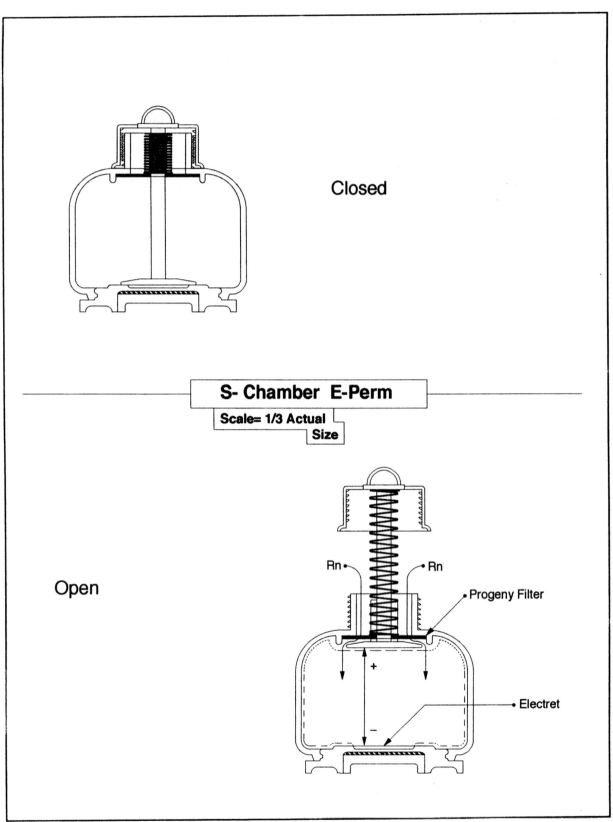

Figure 7 Electret monitor.

Radon concentrations have significant diurnal and seasonal variability; therefore, short-term measurements are not accurate. Tests lasting several months are needed to accurately assess the radon status.

Placement of the monitors is another important sampling variable. As a general rule, canisters should be placed in poorly ventilated areas and in the lowest portions of the building. This will yield "worst case" measurements. As an alternative, monitors could be placed in all rooms.

The average indoor radon level is approximately 1.3 picoCuries/liter, and about 0.4 picoCuries/liter are found outside.

The EPA has set a radon level of 4 picoCuries/liter as the threshold between "safe" and "unsafe" levels. However, health effects on humans exposed to radon levels above 4 picoCuries/liter over long time periods have yet to be documented.

4.3.3 Formaldehyde

Formaldehyde has been used extensively in building materials and furnishings so it is no surprise that the presence of formaldehyde vapors is the most common indoor air pollution problem in residential homes. Building products emitting high levels of formaldehyde include fiberboard, hardwood plywood paneling, particleboard, and urea-formaldehyde foam insulation. Materials emitting lesser amounts of formaldehyde include upholstery fabrics, carpeting, paper products, fiberglass products, sizing agents for clothing, and softwood plywood.

Of these products, urea-formaldehyde foam insulation (UFFI) has received the most attention due to its widespread use as a retrofitted insulation. The Consumer Product Safety Council unsuccessfully attempted to ban this material in the US. Although the use of UFFI has been restricted on a voluntary basis since the mid 1980's, this material is experiencing a small resurgence as an insulating material for concrete blocks and panels used in construction of industrial and commercial buildings. In addition, a UFFI variant is being marketed for use as retrofit residential insulation.

Formaldehyde levels are highest in mobile homes because of the large quantity of formaldehyde sources; average air concentrations commonly range from 90 to 460 ppb. In contrast, conventional homes have substantially less formaldehyde vapors and mean concentrations range from 30 to 140 ppb.

Health effects related to formaldehyde exposure include irritant effects (eye, upper respiratory, and skin), sensitization that causes allergy-like symptoms, and carcinogenicity in laboratory animals. Risk assessments by the EPA have projected that exposure for more than 10 years (such as living in a mobile home) to an average level of 100 ppb may increase cancer by 2 in 10,000.

Sampling is complicated by the fact that environmental factors and decay can greatly affect concentrations. As temperature increases, formaldehyde vapors increase; an approximate 5 degree C increase in temperature often doubles the formaldehyde concentration. An increase in relatively humidity from 30 percent to 70 percent can increase formaldehyde levels approximately 40 percent. Vapor concentrations also change from season to season; levels are highest in the summer and lowest in the winter. Off-gassing of formaldehyde products significantly decreases with time depending on product emission strength, presence of multiple sources, and factors such as local climate and occupant behavior. The most rapid decrease is usually measured during the first year after instillation, but emissions will continue as long as the formaldehyde products are present.

Recommended sampling procedures are as follows:

- close the structure for a minimum of 12 hours prior to sampling and keep the structure closed during sampling;

- maintain a temperature of 25 degrees C (the high end of the normal living range) prior to and during sampling;

- sample during moderate weather, preferably in the summer when relative humidity is less than approximately 60 percent;

- collect a minimum of two samples within the structures using vacuum bottles or pumps;

- sampling duration ranges from 60 to 90 minutes, depending on the expected concentrations; and

- record indoor air temperature, outdoor air temperature, and humidity.

After laboratory determination of the formaldehyde concentration, the use of a standardization equation is recommended because the equation compensates for temperature and humidity.

Formaldehyde measurements are usually taken following peak values that commonly occur within a few weeks to a year after formaldehyde product installation. Therefore, interpretation of test results should include a quantitative assessment of prior exposures using degradation curves and other information.

4.3.4 Biogenic Particles

Particles from biological sources, such as bacteria, fungi, viruses, insect parts and waste, animal saliva, etc. are termed biogenic particles. The relationship between certain biogenic particles and infectious diseases has been known for almost a century. Such diseases are usually outside the scope of indoor air quality studies. However, allergies and asthma are two common diseases that are often within the realm of indoor air quality studies. In particular, the role of dust mites and molds in causing or aggravating allergies and asthma may be part of the indoor air quality program.

One of the most significant causes of asthma and to a lesser extent allergies is the dust mite. These microscopic insects consume sloughed off skin scales from humans and animals; dust mites and their feces are commonly found in household dust. A major limiting factor to their survival is water in the form of humidity. Therefore, these insects are most abundant in humid areas. Sampling for dust mites is still in its infancy. Dust samples are collected by a vacuum device and analyzed for mite allergen by the use of radioimmunoassay methods. Another method detects the guanine (the major excretory products from dust mites) in dust samples and relates this, in a semiquantitative manner, to mite population.

Molds are another significant cause of asthma and allergies. Most mold spores are about 10 microns in size and are produced in large concentrations that remain airborne for long periods of time. These spores can be inhaled and are the major cause of mold allergy. The prevalence of indoor molds is primarily related to humidity and the presence of cooling coil drip pans and other sources of water.

Molds are usually sampled using volumetric air samplers. Air is typically sampled at a high flow rate (1 cubic foot/minute) through a multi-pored orifice plate. Air jets from the pores impinge on a culture medium. Following a 1 to 3 minute sampling period, the plates are closed and incubated. Colonies are then counted and identified. Mold counts above 1000 colony-forming units per cubic meter are considered high while levels below 100 colony-forming units per cubic meter are considered low. This method is limited in that it counts only live or viable mold particles.

Nonviable particles are equally allergenic and are as abundant as the viable forms. When both viable and nonviable molds are to be measured, they can be collected on a glass plate by impaction with a Bukhard sampler and quantified by counting using visible light microscopy.

In addition to sampling for molds, it is important to assess factors that cause water to remain on surfaces and in the air, and to assess the movement from mold infected areas such as basements to living areas.

4.4 SOURCES OF INFORMATION

Godish, T. 1991. Air Quality, second edition. Lewis Publishers, Inc. Chelsea, MI.

Godish, T. 1989. Indoor Air Pollution Control. Lewis Publishers, Inc. Chelsea, MI.

Lesue, G.B. and F.W. Lunau. 1992. Indoor Air Pollution: Problems and Priorities. Cambridge University Press. New York.

Environmental Protection Agency. 1989. Interim Final RCRA Facility Investigation (RFI) Guidance, Air and Surface Water Releases. US EPA. Washington, D.C. EPA 530/SW-89-031.

Environmental Protection Agency. 1989. Model Curriculum for Training Asbestos Abatement Contractors and Supervisors, Section XIV--Sampling and Analytical Methodology Pertaining to Asbestos Abatement. US EPA. Washington, D.C. EPA CX-814627-01-0.

— Unit 5 —
Solids Sampling

Table of Contents

5.1	Regulatory Basis		194
5.2	Representative Sampling		195
5.3	Soil Characterization		196
	5.3.1	Soil Diagnostic Properties	198
5.4	Soil Sampling		203
	5.4.1	Sampling Approach	203
	5.4.2	Sampling Equipment	204
		5.4.2.1 Surface Sampling Tools	204
		5.4.2.2 Sub-surface Sampling Tools	205
	5.4.3	Test Pits and Trenches	208
		5.4.3.1 Sampling Techniques	209
		5.4.3.2 Backfilling	209
5.5	Soil Sediments/Sludges		209
	5.5.1	Sampling Approach	209
	5.5.2	Sampling Equipment	211
5.6	Sampling Sampling Bulk Materials		214
	5.6.1	Sampling Approach	214
	5.6.2	Sampling Equipment	214
5.7	Sampling Artifacts		215
5.8	Sources of Information		215
Appendix			216

— Unit 5 —

SOLIDS SAMPLING

Solids are defined as any material which can be sampled using devices commonly associated with collecting soils, sludges, sediments, bulk materials (such as soil pile, grain, etc.), and solid artifacts (such as batteries, buildings, etc.). Sludges are somewhat ill-defined materials but have properties such as a percent dry weight greater than 10 percent and a viscosity significantly greater than water (non-free flowing media).

Attempting to sample solids in a representative manner can be a very difficult task due to the relatively heterogeneous nature of these materials. Thus, it is essential that other means of preliminary investigations be pursued in order to maximize the information gained from the sampling effort and minimize personal exposure risks. This may include activities such as reviewing historical information and aerial photographs, performing a site reconnaissance, and non-invasive sampling/analysis techniques, etc.

The reactivity of the analyte (contaminate or substance being characterized or analyzed) should be considered when choosing the sampling equipment. Most commercially available solids sampling devices are steel, brass, or plastic. Stainless steel sampling devices are commonly used and are especially appropriate for sampling organics. Use of chrome- or nickel-plated steel should be avoided due to sample contamination potential. High density polyethylene is particularly applicable material for inorganic species.

5.1 REGULATORY BASIS

Hazardous materials have been identified at many thousands of sites across the U.S. Almost all these sites have some sort of solid material (primarily soils) which are contaminated. Soil and sediment contamination are viewed as conduits or pathways through which contaminants pass to other media (such as ground water, atmosphere) which have more direct exposure risks.

The two prominent regulations addressing contaminated solids are RCRA and CERCLA. Additionally, the Department of Transportation (DOT) regulations are pertinent regarding the shipment of hazardous materials. Additional regulations not directly applicable must also be considered due to the indirect exposure risks associated with contaminated solids. For example, contaminated sediments may become soluble which can cause surrounding water to become contaminated. This situation could involve issues associated with the Clean Water Act or Safe Drinking Water Act. It is important to identify which regulations pertain to a particular site investigation due to their impact upon the investigative process. For example, before a material can be sent to a landfill it must be determined whether it is a characteristic waste as defined by RCRA. This entails specific analytical and sampling protocols.

Recently, the U.S. Department of Energy (DOE) has made a major commitment to remediate its nuclear production facilities. These sites can be extremely complicated in terms of the heterogeneous nature of the wastes (such as radionuclides, hazardous wastes, mixed-wastes) and the potential surface extent of the contamination. Basically, the DOE follows the CERCLA process of remediation.

5.2 REPRESENTATIVE SAMPLING

A basic overview of various sampling techniques and a definition of representative sampling was already presented in Unit 1. However, the challenge of obtaining a representative sample for solid materials is complicated due to the heterogeneous nature of the material. Even under natural conditions, soils can be heterogeneous with respect to spatial dimensions. An example would be attempting to define the rate of fluid flowing through a soil which has variable spatial distributions of sub-surface clay layers and particle size distributions. One could attempt to measure the hydraulic conductivity of the soil at various locations but the intervals of field measurements will generally have to be closely spaced to accurately account for this spatial variability. The required sampling intensity should be delineated using geostatisitical analysis in conjunction with the project DQOs. Large scale variability within soils can be reduced by grouping similar soils and sampling within these subunits.

Characterization of "heterogeneous wastes" is generally more complicated than attempting to characterize more uniform materials such as soils or sludges. Heterogeneous wastes include materials such as municipal trash, demolition debris, waste construction materials, containerized wastes, solid wastes from laboratories and manufacturing processes, and post-consumer wastes such as transformers, battery casings, and household items. These materials are difficult to characterize due to heterogeneous composition, and the relatively varied and large particle sizes. The principle difficulty arises in attempting to obtain representative samples of a material with disparate elements. Customary sample segregation, compositing, and homogenization schemes used to characterize waters, soils, or sludges are often completely inappropriate for these materials. Waste particle size frequently poses difficulties. According to standard sampling theory, obtaining a representative sample of varied items in the size range of a centimeter or larger may entail collecting tens or hundreds of pounds of material. Large objects cannot be made to fill standard sample containers without extensive sample preparation which may compromise sample integrity. Few analytical laboratories have the capability of performing leaching tests on raw samples of heterogeneous materials because of the large volumes involved and the difficulty of conserving VOCs. Nor are labs well equipped to reduce samples of large, varied items to the tiny, homogenous units used for analysis. In cases where sample grinding and homogenization are possible, they may be inappropriate. For example, in VOC analysis any attempt to grind the sample exposes the volatile compounds to the atmosphere, which results in lower VOC recoveries.

Due to the difficulties of obtaining representative solids samples it is imperative a project is carefully planned. The DQO's must be clearly identified before planning the sampling event and adjusted as additional information becomes available. Refer to Unit 1 for a detailed discussion of DQO's and its application to the characterization of environmental systems.

Several sampling SOPs along with brief discussions are presented in the Appendix as a supplement to the sampling sections.

5.3 SOIL CHARACTERIZATION

Defining the term "soil" is somewhat dependent upon an individuals perspective. A geologist considers soils as the unconsolidated surface material covering rocks and minerals, while a farmer views soils as a habitat for plants to grow and livestock to graze. A more generalized and descriptive definition would be:

> Mineral and organic materials which occupy parts of the earths surface, are capable of supporting plant and biological life, and have properties that reflect the integrated effect of climate and living matter acting upon parent material, as conditioned by relief, over periods of time.

The soil system is dynamic and relatively heterogeneous. The ability to detect the dynamic nature of soils and its degree of heterogeneity is dependent upon the resolution ability of measurement technique. With this picture of the soil system in mind, it is apparent that any individual measurement is a snapshot of a complicated ecosystem.

The physicochemical properties of soils vary as a function of soil formation factors (climate, organisms, relief, parent material, time) and human influences (such as pollutant release, buildings, etc.). For example, plant growth can have a profound influence on soil development and consequential soil properties. Plants serve as a source of organic matter to soils, which can significantly alter soil properties. An increase in organic matter causes the soil to have an increased water holding capacity, an increased ability to cause organic pollutants to become soluble, increased microbial activities, etc. Individual soil characteristics effect the distribution and migration rates of pollutants within the soil. A relatively sandy soil tends to allow pollutants to move quickly which may translate into contamination of aquifers or receiving waters. Thus, a properly planned soil sampling program can yield data to determine the extent of sub-surface contamination and the potential to migrate towards environmental pathways (e.g. groundwater, plant uptake). This is accomplished through evaluation of the spatial distribution of contaminants, characterization of the physicochemical properties of the soil, and modeling to predict contaminant dispersion or potential contamination within variable conditions.

An "average" soil consists of approximately 47 percent minerals, 3 percent organic matter, 25 percent air space, and 25 percent water. Thus, there are several phases within a soil system. Soil sampling typically involves collection of a bulk sample which disturbs the physical and chemical balance of these phases. There are techniques which focus upon sampling specific phases, such as sampling extractable water (e.g. lysimeters) or sampling the gaseous phase (e.g. soil gas vapor probes). It is important to recognize that the spatial extent of a contaminant is usually distributed across these phases. Depending on the project objectives one may need to characterize this distribution. This requires specific expertise in soil science and requires a combination of techniques which focus on sampling these phases separately and as a whole. Furthermore, there are specific analytical methods which focus on specific portions of the soil system.

5.3.1 Soil Diagnostic Properties

Soils are classified according to certain diagnostic properties which differentiate soils of varying characteristics. These diagnostic properties are used to identify soil horizons, which are collectively referred to as a soil profile (Figure 1). These layered soil horizons possess distinguishable properties such as differing texture, color, and structure. There are three major horizons in a well-developed soil profile. Horizons A and B represent the true soil, while horizon C is the sub-soil or weathered parent material. Below the C horizon is the underlying rock which is referred to as the R horizon. The upper A sub-horizons are characterized as having a relatively large amount of biological activity and accumulation of organic matter. Below this active surface zone, but still within the A horizon, is a zone of leaching (E horizon). This leaching zone is present due to infiltration of precipitation through the soil profile. The water serves to dissolve and translocate both soluble constituents and colloidal particles into the lower sub-soil horizons (B horizons). Figure 1 represents a generalized soil profile. The specific formation of a soil profile is dependent upon a complex process of many interrelated factors, as discussed in the previous section. Different soils have different soil horizons which will effect the transport of pollutants through the soil profile. Thus, it is essential the soil profile is adequately characterized so the sampling strategy takes the layered sub-surface horizons into account. This is the only way to ensure a representative sample has been collected.

Proper characterization of a soil profile requires the expertise of an experienced soil scientist. However, an individual can be trained to provide verification of data found within soil surveys and explain spatial variability which may be important. The characterization technique is similar to the description of borehole drilling completed during ground water well installation. In addition to the field descriptions and measurements, representative samples are collected for more detailed laboratory analysis.

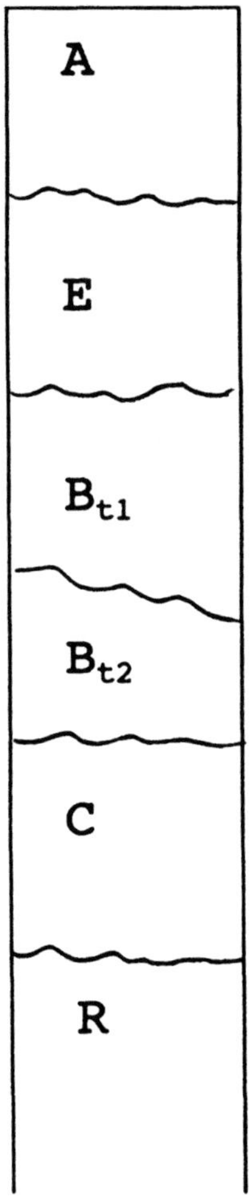

HORIZON	GENERIC HORIZON PROPERTY OR CHARACTERISTICS
A	Surface horizon which has accumulated organic matter and relatively high biological activity
E	Zone of sand and silt particles low in clay, Fe, Al, or organic matter (leached horizon)
B_{t1}	Dominated by obliteration of original rock structure and the accumulation of clay, Fe, Al, organic matter, carbonate, gypsum or Si
B_{t2}	
C	Horizon excluding hard bedrock and little affected by soil formation processess
R	Hard bedrock like basalt, granite or sandstone

The "t" subscript indicates a subordinate horizon in which there is an accumulation of clay. The 1 and 2 designates distinguishable horizons which have some similarities.

Figure 1 Example of a "typical well-developed" soil profile.

The following are commonly used diagnostic properties.

Soil Color

Soil color is the most readily noticed diagnostic property which can be used to infer limited information concerning the development and characteristics of a soil. Soils with a high content of organic matter tend towards brown to black. Color can be useful in identifying layers of poor drainage (they contain orange and reddish spots called mottles). Water-logged, deep soil layers may be light gray (glei or gley) and white soil is sometimes indicative of salt precipitation.

Soil horizons are usually distinguishable by color differences. The Munsell color charts are used to identify soil colors. These charts are used with both dry and wet samples. The recorded value consists of a hue, value, and chroma (such as 7.5YR 4/2).

Soil Texture

Soil texture is one of the most important diagnostic properties. It provides an indication of the proportions of sand, silt and clay, water and gaseous flow, contaminant migration, and plant growth. For example, a clay soil is characterized with a very low hydraulic conductivity (which infers very limited movement of water through the soil profile) whereas a sandy soil would tend to drain water very rapidly.

It is important to note there are a variety of classification systems which define the size limits of a sand, silt, or clay. Within the U.S., three major systems are used: the USDA (U.S. Dept. Agriculture), the USCS (Unified Soil Classification System), and the AASHO (American Association of State Highway Officials). The USDA system is used in soil surveys especially by the agricultural community. However, interpretive information for a wide variety of applications are found within soil surveys. Within the USDA method, the proportions of sand, silt and clay collectively determine the soil texture according to a soil texture triangle (Figure 2). The AASHO is used in highway construction applications. The USCS method is used internationally and has found widespread use in the United States by the Corps of Engineers, the Bureau of Reclamation, and many consulting firms. The USCS is outlined within ASTM method D2487. The differentiation of a silt or clay material is not based upon particle size but instead is defined by two soil engineering tests: plasticity

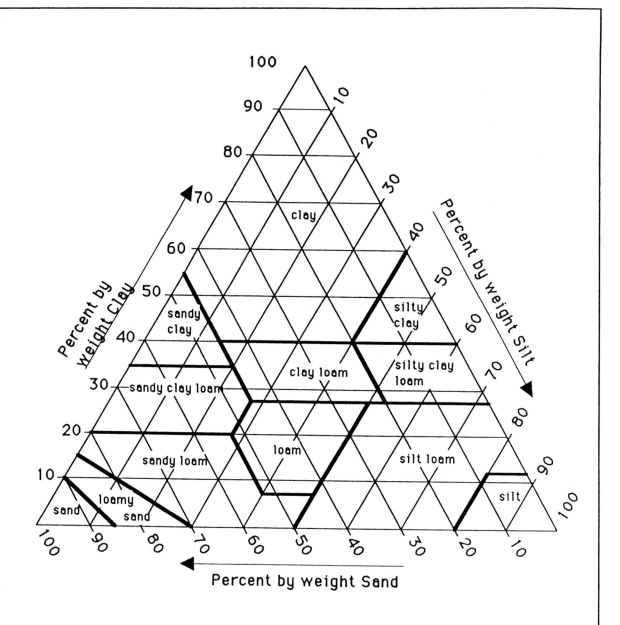

Basic soil textural classes according to the USDA soil classification system as determined by the percentages of clay (< 0.002 mm), silt (0.002 - .05 mm), and sand (0.05 - 2.0 mm). Example: Given 70% silt, 20% sand, and 10% clay, the texture would be identified as a silt loam.

Figure 2 Soil texture triangle.

index and liquid limit. ASTM Method D2487 is applicable for general field descriptions of soils which include properties such as structure, consistency, and color. Soil textures can be identified by "feeling" the soil at its moist-to-wet condition. Two common laboratory procedures used to quantify soil texture include the hydrometer and pipette methods.

Soil Structure

Soil structure is the aggregation of primary soil particles (sand, silt, and clay) into secondary particles (peds). The shape, size, and strength of these peds depend on texture, organic matter, and several other factors. Description of soil structure is completed by comparison of aggregated soil particles to diagrams, and interpretation of the degree of angular nature and elongation. An example of soil structure description is "weak, medium subangular blocky".

Other Physical Properties

A variety of physical measurements are routinely completed as part of general soil characterization. These often include soil moisture content, bulk density, plasticity index, liquid limit, and compaction tests. The soils hydraulic conductivity is important due to its influence on contaminant transport through the soil profile. It should be noted that hydraulic conductivity within the vadose zone (unsaturated zone) is a complicated parameter to characterize due to its dependency upon the soil moisture content and spatial variability. Hydraulic conductivity is defined for saturated flow conditions within the unit on groundwater sampling (Unit 3). It is a parameter which provides an indication of the water percolation through the soil.

Chemical Properties

Common chemical properties measured within general soil characterization includes pH, EC (electrical conductivity), water-soluble and exchangeable cations, cation exchange capacity (CEC), organic matter content, percent lime, and water-soluble anions. Analysis for the contaminant(s) of concern requires considerations such as the expected contaminants at the site, the dominant phases into which the contaminants partition, and environmental factors associated with the site. Selection of chemical properties to measured is dependent primarily upon the project objectives and site specific considerations. Often the investigator will request every conceivable chemical and physical measurement even though much of the data will not help address the issue. Knowing how to properly interpret soils data is essential to efficiently complete an environmental assessment and to ensure the data is of acceptable quality.

5.4 SOIL SAMPLING

5.4.1 Sampling Approach

Soil sampling can be subdivided into surface and sub-surface samples. A surface sample is often defined as having a depth of 0-12" while a sub-surface sample has a depth of greater than 12". Surface sampling is often completed to identify areas where spills and leaks may have occurred, or where the sub-surface transport of the contaminant is negligible, and/or as part of the overall site characterization. Surface contamination may be identified visually through identification of discolored soils or areas with distressed/unusual vegetation appearance. Sometimes, surface contamination can initially be identified and delineated with applicable surface field portable analytical instrumentation (such as total organic vapor analyzer, x-ray fluorescence, or radiation detectors). This can help focus the sampling effort.

Sub-surface sampling will provide an indication of the extent of sub-surface contamination and can provide a more complete characterization of a site when used in conjunction with surface sampling. Sub-surface sampling can be extremely complicated and costly due to the difficulty in obtaining representative samples as sampling depth increases. As previously discussed, the characteristics of soil horizons varies spatially. Representative sampling of these spatially variable horizons is difficult and must be addressed through careful evaluations of diagnostic soil properties. There are also safety concerns associated with encountering buried tanks, drums, pipes, etc. Thus, it is extremely important to sufficiently characterize the sub-surface in a remote manner. This can be accomplished through various geophysical techniques and by completing a thorough preliminary (Phase I) investigation. Sub-surface sampling can be accomplished through the use of specific tools (such as drill augers, tube sampler, etc.) designed to allow penetration of the subsoil, or a test pit (square-like shape) or trench (longitudinal) can be opened.

The specific sampling approach is dependent upon site specific factors and project DQOs. When choosing a specific sampling approach keep in mind that individual sampling points cannot always be regarded as statistically independent, which violates a basic assumption of classical statistics. Thus, evaluation of the sampling effort from a geostatistical point of view should be considered.

5.4.2 Sampling Equipment

Factors which affect the selection of the proper soil sampling tools include project objectives (DQOs), the physicochemical nature of the contamination and its reactivity with soil, soil density, soil texture, soil structure, water content, depth of sampling interval, sample size requirement, site accessibility, past data comparisons, and representativeness of sample (such as soil volume and interval of interest). Due to the number of field variables which will affect the suitability of a sampling tool to collect an acceptable sample, the field sampling crew must have access to a variety of tools. Changes in the sampling plan with regard to the preferred sampling method must be clearly documented and justified before sampling proceeds. Due to spatial variation, even the orientation of a sampling device can have an influence on analytical concentrations. Therefore, it is important to be consistent in the sampling devices used to collect a sample to ensure comparable data.

5.4.2.1 Surface Sampling Tools

Collection of samples from near the soil surface can be accomplished using tools such as spades, shovels, and scoops. These tools have the advantage of being portable, they can be used with a variety of soil conditions, and they are relatively simple to operate. However, it is difficult to obtain a consistent sample with respect to spatial distribution using a shovel or scoop. In addition, there is a high probability of obtaining erroneously low results when sampling for VOCs. This is due to the excessive atmospheric-soil surface area exposed during sample collection.

A thin-wall tube sampler (may also be referred to as core sampler) will obtain more reproducible samples and is better suited for sampling VOCs. This sampler is pushed into the soil to a desired depth and then pulled out of the ground for sample removal. The hand pushed samplers are generally restricted to surface sampling due to the difficulty in penetrating the subsurface. Sampling tube inserts are available which are placed within the tube before insertion of the core into the soil. After the sample has been brought to the surface, the inserted tube is removed and capped off (and sometimes sealed with parrafin). This minimizes the soil exposure to the atmosphere which reduces the degree of analyte volatilization. The sample is now ready for shipment to the laboratory or on-site analysis. The only major drawback to using sampling tubes is the lack of ability to visually inspect the sample. However, clear sampling tubes are available which allow limited visual inspection of the soil. Selection of the appropriate sampling tube insert must consider reactivity of the analytes with the tube material. For example, the use of plastic sampling tube inserts is generally not recommended for organics although a teflon liner would work well. Core samplers do not work well in gravelly and/or very dry, hard soil conditions. Hard soil conditions can still be sampled if a weighted drive force or hydraulic driver is utilized. The ability to penetrate soils is highly dependent upon the soil texture, moisture content, and amount of gravel or rocky materials.

A variety of auger-type samplers can also be used to collect surface samples (Figure 3). Auger samplers generally work better than core samplers in gravelly type soils. However, because of the twisting nature of these samplers the collected sample is more disturbed.

5.4.2.2 Sub-surface Sampling Tools

A variety of soil sampling tools exist which are applicable for obtaining samples at depths greater than 1 foot. Most of them can be used to sample at depth increments of 0 to 6 inches, but are problematic when sampling at shallower depths. The tools can be categorized as screw or barrel augers and tube samplers. These can vary in complexity from simple hand-operated devices to portable, motorized devices to major drilling rig machines used for geologic sub-surface operations.

Tube samplers (thin-walled tube or split barrel sampler) generally yield the best quality sample for characterization purposes. The previous discussion illustrated some of the advantages and disadvantages of tube samplers. It is desirable to have interchangeable cutting tips which allow smooth penetration into the soil with reduced sample disturbance. The sample recovery volume (percentage of coring device occupied by soil) and the degree of resistance to soil penetration (such as the number of blows it takes to drive a sampler into the soil) should be noted within the sample logbook. ASTM-D1586 is a standard penetration

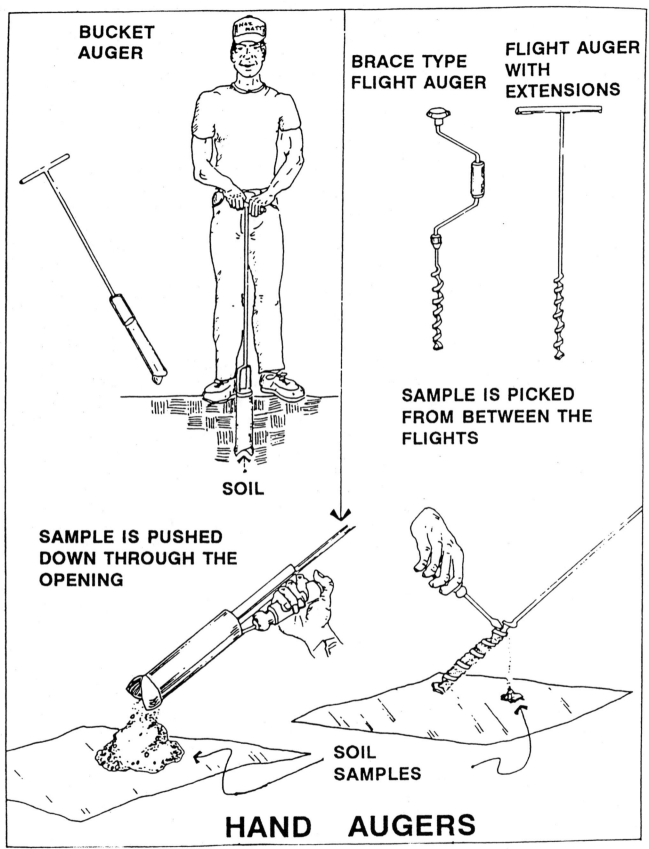

Figure 3 Hand-held soil auger samplers.

method.

Screw augers can be used as part of well-monitoring installation, deep geologic investigations, or for shallow soil sampling. However, the quality of the obtained sample is generally quite poor. A high possibility of cross contamination between soil layers exists because the sample passes through the soil profile in contact with the soil. Also, this leads to difficulty in quantifying the depth from which the sample is obtained (i.e. precise location of changes in soil strata). Other common problems include potential contamination from drilling material and a loss of VOCs. Auger borings are generally applicable when there is no need for undisturbed samples and/or when there is a need for an open borehole which could be used for further diagnostics (such as sub-surface radiation monitoring).

As discussed in Unit 3, a common sampling technique is to drill to the desired sampling depth with an auger. Then, the sample is collected by insertion of a tube sampler. This sequence is repeated until the desired sampling profile has been completed. An SOP for hand-driven sampling is provided in the Appendix. Various powered augers suitable for ground water well construction are discussed in Unit 3. Several forms of augers and tube probes exist which are powered by hand-held hammer drills or are hydraulic units mounted onto vehicles. These highly portable power driven soil samplers can also be used to insert soil/gas sampling probes. Their major limitation relates to their inability to penetrate hard subsoil or rocky conditions to any significant depth, and they are limited to approximately a 20 foot depth even under ideal conditions.

A variety of hand-driven augers exist (Figure 3), which possess unique advantages related to encountered soil conditions. Generally, a hand auger is attached to the bottom of an extension rod that has a crossarm at the top. The hole is drilled by turning this crossarm at the same time the operator presses the auger into the ground. As the auger is advanced and becomes filled with soil, it is unscrewed from the hole, and the soil is removed. Hand-operated augers work well for sampling a wide variety of soils, with the exception of cohesionless material (such as sands). When the soil conditions are excessively hard, then some type of power driven auger is more appropriate.

5.4.3 Test Pits and Trenches

Several advantages associated with open test pits include the ability to accurately characterize the soil profile, increased access to a larger area of soil when compared to a single soil boring, and a more accurate approach to characterizing landfills and dumping areas. Limitations include the handling/disposal of contaminated soils, safety hazards, and soil/ground water conditions which may limit the depth of excavations.

Test pits typically have a cross section of 4 to 10 feet square while test trenches are around 3 to 6 feet wide with variable lengths. Depth of excavations are generally less than 15 feet, with a limitation of several feet below the water table superimposed. Sometimes a pumping system can be used to control water levels within the pits, but then the pumped water must be handled appropriately.

There are a number of health and safety concerns associated with the excavation of trenches at hazardous waste sites. All excavations deeper than 4 feet must be stabilized by bracing the pit sides using wooden or steel support structures, ladders must be in the hole at all times, and a temporary guardrail must be placed along the surface of the hole before entry (refer to OSHA: 29 CFR 1926, 29 CFR 1910.120, and 29 CFR 1910.134). Air monitoring of the excavation is required before entry (percent O_2 and toxic/explosive gases) and appropriate PPE is required. At least two persons should be present before entry by one of the investigators. Entry into the pit by placing a person into a backhoe scoop is not permitted. Other safety and logistical considerations include: 1) allowing sufficient space between excavations, 2) excavated soils should be stock piled to one side, preferably down wind and away from the immediate edge of the pit, 3) control of the overland flow of water and erosion of any stockpiled soil, and 4) the need for a temporary detention basin and drainage system to prevent contaminant migrations.

Generally a test pit log is completed which identifies the soil profile. This should include a sketch of pit conditions, photographs with a scale for reference, and a precise definition of the pit location. Field logbook data may typically include: job ID and location, date of excavation, approximate surface elevation, depth of excavation, pit dimensions, sampling method, soil/rock descriptions, photographs, groundwater level, on-site analytical data (such as air monitoring, soil/gas analysis, etc.), and visual observations (such as dark-stained soil).

5.4.3.1 Sampling Techniques

Determination of sampling methodology must consider site-specific factors and the project DQO's. One approach commonly used is to sample according to the soil horizons. Utilization of the sampling tools already discussed in previous sections is common. Disturbed sampling techniques commonly include sampling the walls or floors of the test pit by means of scraping or digging with a shovel, scoop, or rockpick. Samples are sometimes taken directly from the backhoe bucket. However, this sampling technique does not allow for precise identification of the sample location.

Relatively undisturbed samples can be obtained by isolating a large cube of soil at the base or side of the pit. The sample is cut using knives and shovels taking care to minimize soil disturbance. The entire block of soil is removed and placed in an airtight container for shipment to the lab. Another technique is to insert a core sampling tube into an undisturbed portion of the pit.

5.4.3.2 Backfilling

Backfilling of excavations may require specific compaction standards. The soil horizons may need to be reconstructed, especially if a relatively low permeable layer was penetrated. In this case, the backfill could consist of a soil-bentonite mix in a proportion which creates a permeability equal to or less than the original conditions. The area may also need to be revegetated to minimize soil erosion and maintain consistency in the land use.

5.5 SAMPLING SEDIMENTS/SLUDGES

Sampling sediments and sludges are discussed together because often there is no distinction between these materials and the sampling tools utilized are interchangeable. The health and safety concerns of sampling hazardous sludges are not addressed within this unit but should be recognized and dealt with appropriately. This discussion is directed primarily towards open systems (i.e. non-containerized).

5.5.1 Sampling Approach

Sampling sediments or sludges requires extensive pre-planning activities such that the project DQOs can be met in a cost effective manner. This is particularly relevant due to the unique complexities associated with attempting to sample sediments. For example, representative sediment sampling of a lake will usually require the use of a boat. This increases the likelihood of potential safety concerns, makes it more difficult to use the sampling equipment, and further complicates sample handling and record keeping. Additionally, the sediments are generally not visible and the available sampling equipment has limitations which makes it difficult to collect a representative sample. Another example would be attempting to obtain a sediment sample from the middle of a shallow, rapidly moving stream. This scenario would preclude the use of a boat, and walking out into the stream may be too hazardous and/or may cause sample bias. One approach to sampling may be to support a wire at each side of the stream bank and to remotely obtain a sample using the wire, a sampling device and remote robotic equipment.

Of course, the sampling locations are highly dependent upon the study objectives. Often sediments are sampled to examine the degree to which contaminants may concentrate along a water-bearing bottom relative to a waste source. This situation would require sampling upstream from the waste discharge source to establish a background concentration, and then sampling at intervals downstream from the source. Other common approaches to selecting sampling locations include: 1) sample where there is visible contamination, 2) sample at the same location where a water/liquid sample is collected, 3) systematic sampling.

Streams, lakes, and impoundments may possess significant variations in sediment/sludge composition with respect to distance from inflows, discharges, or other disturbances. This makes it important that the sampling location is located precisely, which may involve triangulation with stable references on the banks of the water-bearing zone or the use of more sophisticated Global Positioning System (GPS) instrumentation. This potential spatial variability not only extends horizontally, but also vertically. The degree of variability should be assessed before designing the sampling plan because it will have a large influence on the number and location of sampling points, and the appropriate sampling equipment. The heterogeneity of the material may be reduced through bias selection of the sampling points. For example, try to select locations where there are observable differences in sediment type or depth.

Sampling equipment should be selected depending on the physical nature of the sediment material. Material such as rocks, organic matter, and artifacts can preclude the use of or require modification to sampling devices. However, sediments can be viewed as a mixture of soil, water, organic matter, rocks, plants and even trash. Characterization may necessitate representative sampling in each of these phases which is a very difficult task. Sedminents may be more like a sludge (de-watered solids to highly viscous liquids) which possess a low specific density or it may be cohesiveless (loose) material which could potentially fall out of a sampling device.

Remember, if the sampling scheme requires water and sediment/sludge sampling, sample the water first. Sediment/sludge sampling has a high potential for contaminating the water.

Attempting to sample sludges in a representative manner can present some unique challenges. Depending upon the formation process of the sludge (such as industrial source, wastewater treatment plant, environmentally aged sludge, etc.) and the project objectives, the sampling approach will be different. For example, a representative sample may be obtained by collecting two grab samples using a scoop attached to an extended rod when the sludge is a well-mixed, relatively homogenous media. This type of sludge can be typical of an industrial process. However, if that same sludge were dumped into a ditch which intermittently had standing water and aged for several years then the material is probably going to be quite heterogeneous. Obtaining a representative sample from this heterogeneous material is difficult and requires a carefully planned approach.

5.5.2 Sampling Equipment

A variety of sampling equipment is available for sediment/sludge sampling. The previous discussions illustrated that a variety of equipment should be available and alternative plans developed depending on the specific conditions encountered in the field. If the sample has the potential for being considered a hazardous material, then disposable sampling equipment should be considered or adequate decontamination procedures employed and proper safety measures must be developed. It must be decided whether the sample must represent the vertical profile of the sediment/sludge or whether a disturbed, surface sampling is adequate. The physical nature of the material being sampled must also be considered in equipment selection. Additionally, it is important to consider the reactivity of the compounds of interest.

Disturbed samples can be collected using a dredge bottom grab sampler (Ekman & Ponar grab samplers), a modified auger, or even a simple scoop or dipper may be suitable. Scoops (miniature shovels) are limited to conditions of very shallow water or materials having a low water content (SOP in the Appendix). The Ponar grab sampler operates by lowering the sampler through the water with a line attached for sample retrieval (Figure 4) and the sampler is driven into the solids material as it impacts the sediment. The tension associated with lifting the cable activates the lever system to close the clamshell, which holds the disturbed sample in place (SOP in the Appendix). However, the triggering device has been noted to work improperly on some models. The Ponar sampler is designed for taking samples of hard bottoms such as sands, gravels, or clays. Another widely used grab sampler is the Ekman sampler. It works similarly to the Ponar sampler except a messenger is sent down the line which strikes the release mechanism thereby releasing two hooked cables which allows the spring loaded jaws to close. The Ekman sampler can also be used with an extension handle which permits easier shallow water sampling operations. The Ekman grab samplers are designed to collect samples in soft, finely divided muck, mud, ooze, or fine peaty materials which are free from vegetation, and intermixtures of sand, stones, or other coarse debris. Due to the weight (up to 60 pounds empty) and depths to which these grab samplers must reach, one normally uses medium size boats which are equipped with winches so the grab samplers can be readily deployed and retrieved. Petite size versions are available in which hand operation is possible.

A dipper sampling cup located on the end of an extension pole (normally used for water sampling) can be adapted to sample sludges of high water content. It is wise to use the actual laboratory sample bottle container as the sampling cup so that there is minimal sample handling and decontamination between sampling is not an issue of concern.

Relatively undisturbed samples can be obtained using a variety of core samplers. The sampled media is generally retained within the cylinder using a retainer at the lower end and/or a check valve at the upper end (Figure 4). The sediment core samplers may have fins and/or weights to help guide it through the water. Specific core sludge sampling devices exist which generally have a core retainer at the lower end. The ability of these sludge samplers to

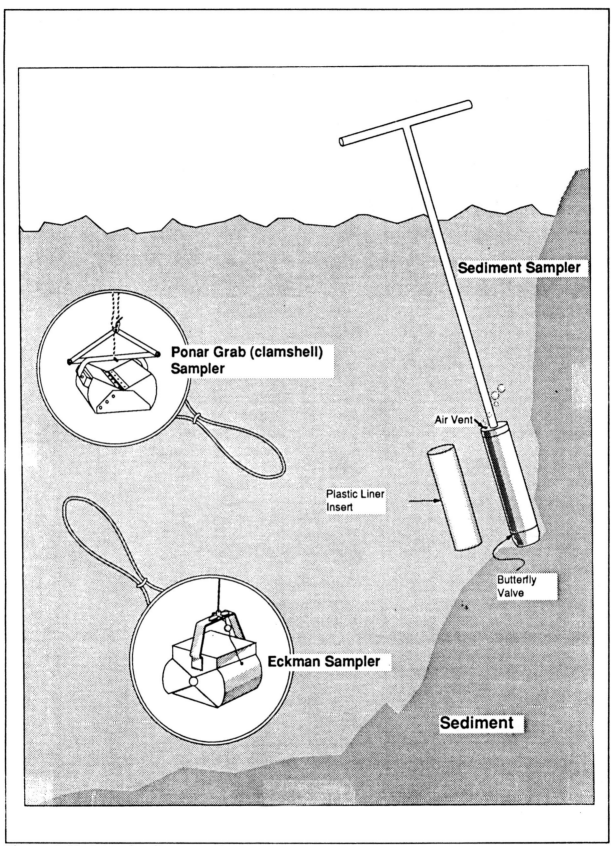

Figure 4 Variety of sediment tools.

effectively operate is somewhat dependent upon the physical properties of the sludge. A modified soil bucket auger may also be used with some success. Several SOPs are presented in the Appendix.

5.6 SAMPLING BULK MATERIALS

5.6.1 Sampling Approach

Bulk materials are often a relatively homogenous collection of a single identifiable product. These materials are generally containerized, although the material may be pile directly on the ground. The heterogeneity present within an exposed pile can be predicted based upon the degree of surface exposure to the atmosphere and/or background information on the manufacturing or sources of the material. Thus, one must consider that there can be changes in composition with respect to time (degree of exposure to the atmosphere or degree of "maturation") and with respect to distance (changes in the composition across the cross-section of the pile). This requires collecting grab samples which will characterize this structured heterogeneity. Generally the sampling approach will include representative sampling of the outer surface zone and then the interior portions of the pile. A coring device would facilitate this type of sampling. If there is no reason to expect such heterogeneity, then it is common to composite a number of grab samples in order to increase the representativeness of the analyzed sample and reduce analytical costs. A composite sampling approach should be pre-planned to insure that the resulting data is meaningful.

5.6.2 Sampling Equipment

Bulk materials in an unconsolidated state may be readily collected by a stainless steel/polyethylene scoop, a coring device with a sample catcher at the end, a trier or a grain thief. The grain thief is limited to unconsolidated materials which possess particle sizes smaller than the openings. A grain thief consists of a long hollow tube with evenly spaced openings along its length. The tube openings are initially in a closed position via an outer sleeve as the thief is inserted into the pile. Upon complete insertion, the inner sleeve is rotated until its openings align with those on the outer sleeve. The bulk material then enters the device, the inner sleeve is rotated to a sealed position, and the device is withdrawn (SOP in the Appendix).

5.7 SAMPLING ARTIFACTS

Attempting to characterize potential contamination of artifactual materials, such as a wall within a building or a counter-top, can be quite challenging. Obtaining a representative cross-sectional sample from a wall or a monolithic structure would require using a powered coring device. Such activity requires careful planning to ensure sample integrity and that the operation was completed in a safe manner. Reduction of the sample for laboratory analysis is very challenging, especially for VOAs. The results of TCLP analysis will vary greatly depending on the degree to which particle size reduction is employed.

Determination of surface contamination can be completed easily using a variety of field screening instruments and/or by completing a wipe sample. Wipe sampling is also used to ascertain whether equipment decontamination has been effective. Generally, wipe sampling entails rubbing a moistened filter paper over a measured area of 100 cm^2 to 1 m^2 and the paper is then sent to the laboratory for analysis. It is important to quantify the exact surface area wiped as the analytical results must be related to the known area of the sample. The sampler should use a clean, impervious disposable surgeon's glove for each sampling. The filter is moistened with a solvent which will dissolve the contaminants of concern. Typical solvents include distilled, deionized water or the solvent used in the lab analysis. An informal SOP is presented at the end of the Appendix.

5.8 SOURCES OF INFORMATION

Environmental Protection Agency. 1984. Characterization of Hazardous Waste Sites-A Methods Manual: Vol. II. Available Sampling Methods. EPA/600/4-84-076.

Environmental Protection Agency. 1992. Characterizing Heterogeneous Wastes: Methods and Recommendations. EPA/600/R-92/033.

Environmental Protection Agency. 1987. A Compendium of Superfund Field Operation Methods. EPA/540/P-87/001.

Singer, M. and D. Munns. 1991. Soils: An Introduction. Macmillan Publishing Company, N.Y.

APPENDIX

REPRESENTATIVE SOLIDS SAMPLING

STANDARD OPERATING PROCEDURES (SOPs)

The following SOPs are presented to provide an introduction to some additional details regarding a selected number of sampling procedures for solid materials and to provide an illustration of an SOP. The SOPs possess variable levels of detail which provides an understanding that not all SOPs include every required step in a sampling effort. These SOPs should not be interpreted as being correct for a specific sampling program as there may be site specific considerations and project objectives which are not accounted for in the presented procedures. Additionally, these SOPs do not specify all the required levels of detail. For example, issues such as documentation of sampling locations and decontamination procedures are not addressed. Unless otherwise noted, the SOPs were obtained from the following document:

Environmental Protection Agency. 1984. Characterization of Hazardous Waste Sites-A Methods Manual: Vol. II. Available Sampling Methods. EPA/600/4-84-076.

SOIL SAMPLING WITH A SPADE OR SCOOP

Discussion

One of simplest methods of collecting soil samples is with a spade or scoop. A normal lawn or garden spade can be utilized to remove the top cover of soil to the required depth and then a smaller stainless steel scoop can be used to collect the sample.

Uses

This method can be used in most soil types but is limited to sampling the near surface. Samples from depths greater than 12 inches become extremely labor intensive in most soils. Representative samples can be collected with this procedure depending on the care and precision employed by the sampler. The use of a flat shovel to cut a block of soil can be used to obtain less disturbed soil profiles. A stainless steel scoop will suffice for most other applications. Care should be exercised to avoid the use of devices plated with chrome or other plating materials, which is common in garden tools.

Procedures for Use

1. Carefully remove the top layer of soil to the desired sample depth with a pre-cleaned spade.

2. Using a pre-cleaned stainless steel scoop or trowel, remove and discard the thin layer of soil which was in contact with the shovel, and collect a representative sample. If compositing a series of grab samples, use a stainless steel mixing bowl or Teflon tray for mixing. (NOTE: If a pre-cleaned stainless steel spade was used, the removal of the thin soil layer is not necessary. Sampling directly with this type of spade is possible.)

3. Transfer sample into an appropriate sample bottle with a stainless steel lab spoon or equivalent.

4. Check that a Teflon liner is present in the cap if required. Secure the cap tightly. The chemical preservation of solids is generally not recommended, but refer to the Sampling and Analysis Plan and/or check with the laboratory. Refrigeration is usually the best approach supplemented by a minimal holding time.

5. Label the sample bottle with the appropriate sample tag. Be sure to label the tag carefully and clearly, addressing all the categories or parameters. Complete all chain-of-custody documents and record in the field log book.

6. Decontaminate equipment after use and between sample locations.

SUB-SURFACE SOLID SAMPLING WITH AUGER AND THIN-WALL TUBE SAMPLER

Discussion

This system may consist of an auger bit, a series of drill rods, a "T" handle, and a thin-wall tube corer. The auger bit is used to bore a hole to the desired sampling depth and is then withdrawn. The auger tip is then replaced with the tube corer, lowered down the borehole, and forced into the soil at the completion depth. The corer is then withdrawn and the sample is collected.

Alternatively the sample can be recovered directly from the auger. This technique does not provide an "undisturbed" sample as would be collected with a thin tube sampler. In situations where the soil is rocky, it may not be possible to force a thin tube sampler through the soil or sample recovery may be poor. Sampling directly from the auger may be the only viable method. Several auger types are available which include the bucket type and continuous flight (screw). Bucket types are good for direct sample recovery. They are fast and provide a large volume of sample. When screw augers are utilized, the sample is collected directly off the flights, however, this technique will provide a somewhat unrepresentative sample as the exact depth will not be known. The screw auger may be satisfactory when a composite of the entire soil column is desired. In soils where the borehole will not remain open when the tool is removed, a temporary casing can be used until the desired sampling depth is reached.

Uses

This system can be used in a wide variety of soil conditions. It can be used to sample both from the surface by simply driving the corer without preliminary boring, or to depths in excess of 10 meters when there are no rocks present or tendencies of the borehole to collapse.

Procedure for Use

1. Clear the area to be sampled of any surface debris. It may be advisable to remove the first 8 to 15 cm of surface soil for an area approximately 15 cm in radius around the drilling location.

2. Begin drilling, periodically removing accumulated soils. This prevents accidentally brushing loose material back down the borehole when removing the auger or adding drill rods.

3. After reaching desired depth, slowly and carefully remove the auger from the boring. (Note: When sampling directly from the auger, collect the sample after the auger is removed from the boring and proceed to Step 7.)

4. Carefully lower corer down borehole. Gradually push corer into soil, taking care to avoid scraping the borehole sides.

5. Carefully remove corer from borehole.

6. Discard the top of the core (~ 2.5 cm) which represents material collected by the corer before penetration of the layer being sampled. Place the remaining core into a sample container.

7. Check that a Teflon liner is present in the cap if required. Secure the cap tightly. The chemical preservation of solids is generally not recommended, but refer to the Sampling and Analysis Plan and/or check with the laboratory. Refrigeration is usually the best approach supplemented by a minimal holding time.

8. Label the sample bottle with the appropriate sample tag. Be sure to label the tag carefully and clearly, addressing all the categories or parameters. Complete all chain-of-custody documents and record in the field log book.

9. Decontaminate equipment after use and between sample locations.

SAMPLING SOILS, SEDIMENTS, OR SLUDGES WITH A HAND CORER

Discussion

Sampling soils with a hand coring device can be completed as the previously described auger and corer SOP or with the hand corer alone. In the case of sediment or sludge sampling, the hand corer is modified by the addition of a check valve on the top and/or a core sample catcher at the bottom to prevent washout during sample retrieval.

Uses

Hand corers have the advantage of collecting relatively undisturbed samples which can reflect profile stratification in the sample. Extension handles will allow collection of samples underlying a shallow layer of liquid or deeper sampling efforts. Most corers facilitate liners which are available in brass, polycarbonate plastic or Teflon. Care should be taken to choose a material which will not compromise the intended analytical procedures.

Procedures for Use

1. Inspect the corer for proper precleaning and core liner. Select sample location and note the location in field notes.

2. Push corer in with a smooth continuous motion until the sampling depth has been reached.

3. Twist corer and then withdraw in a single smooth motion.

4. Withdraw sample from corer into a stainless steel or Teflon tray. (NOTE: Depending on the sampling device used, additional or alternative steps may be needed here. For example, use of a core liner would alter this step to the following: Remove core liner from the core sampling device. Seal ends of core liner with sample end caps. Proceed to Step 6 but replace "sample bottle" with "sample core".)

5. Transfer sample into an appropriate sample container with a stainless steel lab spoon or equivalent.

6. Check that a Teflon liner is present in the cap if required. Secure the cap tightly. The chemical preservation of solids is generally not recommended, but refer to the Sampling and Analysis Plan and/or check with the laboratory. Refrigeration is usually the best approach supplemented by a minimal holding time.

7. Label the sample bottle with the appropriate sample tag. Be sure to label the tag carefully and clearly addressing all the categories or parameters. Complete all chain-of-custody documents and record in the field log book.

8. Decontaminate the equipment.

SAMPLING BOTTOM SLUDGES OR SEDIMENTS WITH A GRAVITY CORER

Discussion

A gravity corer is a metal tube with a replacement tapered nosepiece on the bottom and a ball or other type of check valve on top and/or a sample core catcher at the bottom. The check valve allows water to pass through the corer on descent but prevents a washout during recovery. The tapered nosepiece facilitates cutting and reduces core disturbance during penetration. Most corers are constructed of brass or steel and many can accept liners and additional weights.

Uses

Corers are capable of collecting samples of most sludges and sediments. The gravity corers are well suited for sampling sediments and sludges which are overlaid with at least 5 feet of water. Depending on the density of the substrate and the weight of the corer, penetration to depths of 75 cm can be attained.

Care should be exercised when using gravity corers in vessels or lagoons that have liners because penetration depths could exceed that of the substrate and result in damage to the liner material.

Procedure for Use

1. Attach a pre-cleaned corer to the required length of sample line. Solid braided 5 mm (3/16 inch) nylon line is sufficient; 20 mm (3/4 inch) nylon is easier to grasp during hand hoisting.

2. Secure the free end of the line to a fixed support to prevent accidental loss of the corer.

3. Allow corer to free fall through the liquid to the bottom.

4. Retrieve corer with a smooth, continuous lifting motion. Do not bump the corer as this may result in some sample loss.

5. Remove nosepiece from corer and slide sample out of corer into stainless steel or Teflon pan if a liner is not used. (If a liner is used else go to Step 7)

6. Transfer the sample into an appropriate sample container with a stainless steel lab spoon or equivalent.

7. Check that a Teflon liner is present in the cap if required. Secure the cap tightly. The chemical preservation of solids is generally not recommended, but refer to the Sampling and Analysis Plan and/or check with the laboratory. Refrigeration is usually the best approach supplemented by a minimal holding time.

8. Label the sample bottle with the appropriate sample tag. Be sure to label the tag carefully and clearly, addressing all the categories or parameters. Complete all chain-of-custody documents and record in the field log book.

9. Decontaminate equipment after use and between sample locations.

WIPE SAMPLING

The following are steps involved in obtaining a wipe sample. This information was based on the following document: Environmental Protection Agency. 1987. A Compendium of Superfund Field Operations Methods. EPA/540/0-87/001. This is not a formalized SOP.

1. Using a clean, impervious disposable glove, such as a surgeon's glove, remove a filter paper from the box. (**Note:** Although it is necessary to change the glove if it touches the surface being wiped, a new glove should be used for each sample to avoid cross contamination of samples.)

2. Moisten the filter with a collection medium selected to dissolve the contaminants of concern as specified in the sampling plan. Typical solvents include organic-free water or the solvent used in analysis. The filter should be wet but not dripping.

3. Thoroughly wipe a specified area with the moistened filter. It is important that the area wiped is precisely controlled; the use of a templet may assist in controlling the area wiped. If a different size area is wiped, record the change in the field logbook. If the surface is not flat, be sure to wipe any crevices or depressions and record this in the logbook.

4. Without allowing the filter to contact any other surface, fold it with the exposed side in, and then fold it over to form a 90-degree angle in the center of the filter.

5. Place the filter (angle first) into a clean glass jar, replace the top, seal the jar according to QA requirements, and send the sample to the appropriate laboratory.

6. Prepare a blank by moistening a filter with the collection medium. Place the blank in a separate jar and submit it with the other samples.

7. Document the sample collection in the field logbook and on appropriate forms, and ship samples per specified procedures.